UNITED STATES CONTRIBUTIONS TO QUATERNARY RESEARCH

*The printing of this volume
has been made possible
through the bequest of
Richard Alexander Fullerton Penrose, Jr.
and is partially supported by
The National Science Foundation.*

United States Contributions to Quaternary Research

ATUETTE DE BRASSEMPOUY

Papers Prepared on the
Occasion of the
VIII Congress of the
International Association for
Quaternary Research
Paris, France, 1969

Edited by
STANLEY A. SCHUMM
Colorado State University
Fort Collins, Colorado

WILLIAM C. BRADLEY
University of Colorado
Boulder, Colorado

THE GEOLOGICAL SOCIETY
OF AMERICA, INC.
BOULDER, COLORADO
1969

**Special Paper
Number 123**

Published by
THE GEOLOGICAL SOCIETY OF AMERICA, INC.
P.O. Box 1719
Boulder, Colorado 80302

Printed in the United States of America

Contents

v

Preface

At the 1968 meeting of the U.S. National Committee for INQUA (International Union for Quaternary Research) there was agreement that an effort should be made to induce the editors of various scientific journals to publish special INQUA issues as an American contribution to the VIII INQUA Congress. The editor of the Geological Society of America publications, Edwin B. Eckel, committed his organization to such a publication; however, rather than impose a special INQUA issue of the Bulletin on the membership, a volume in the Special Paper series was proposed. This seemed appropriate to all concerned, but in view of the limitations of time the acquisition and publication of invited papers would have been difficult. It was then suggested that rather than attempting to attract or to invite papers for publication a more representative American contribution would be obtained if those manuscripts that were being evaluated for publication and those that would be received in the normal course of events during the forthcoming months were published in the Special Paper. The papers published in this volume, therefore, are those that would have been published during a period of about a year in the Bulletin. They are a sample taken during 1968 and 1969 of the type of Quaternary research carried out in this country, and the volume is offered as such as an American contribution to the VIII INQUA Congress.

It is worth noting that the papers are in several instances the results of Ph.D. research, providing an encouraging preview of future efforts by these young men and the future status of Quaternary studies in the United States.

The collection would appear not to be representative in terms of subject material with three of the eleven papers dealing with some aspect of landslides or avalanches. One of these papers is of especial interest because the author attempts to apply the results of a Quaternary study to engineering geology problems. This is a promising development, for at a time when great concern is being shown with regard to environment changes, the Quaternary scientist should be aware that he is a specialist in the study of paleoenvironmental changes and that he may be able to contribute significantly to modern environmental problems.

STANLEY A. SCHUMM
WILLIAM C. BRADLEY

GEOLOGICAL SOCIETY OF AMERICA, INC.
SPECIAL PAPER 123

Holocene Climatic and Glacial History of the Central Sierra Nevada, California

ROBERT R. CURRY

University of California, Santa Barbara, California

ABSTRACT

Climatic data for the Sierra Nevada are compared with the range of climatic conditions that are inferred to have occurred during the last 10,000 years. A model of paleoclimatic fluctuation based chiefly upon variation in mean seasonal snowfall is derived. Using a model based upon fluctuations in mean snowfall of less than 50 percent of the mean snowfall during the standard climatic normal period (1931–1960), I propose a chronology of climatic fluctuations for the Sierra Nevada for the last 10,000 years. This chronology is based upon paleoclimatic data derived from study of geologic deposits dated by radiocarbon and lichenometry, variation in tree rings, changes in timberline position, vegetational age classes, and direct and indirect historical records.

At least four major periods of increased mean snowfall and cooler, cloudier summers during the last 10,000 years resulted in four periods of multiple glacial advance in the Sierra Nevada. These occurred between 6000 and 7000 years ago, between 2000 and 2600 years ago, around 1000 years ago, and between 650 years ago and the present. The latest major period of net accumulation and advance in all cirques that are presently occupied by residual glaciers occurred between 1880 and 1908 with a peak from 1895 to 1897.

The fluctuations from glacial to interglacial climate that are presently known to have occurred during the last 8000 years can be fully explained by a climatic model in which the extremes of mean precipitation for the 96 years of historical record are greater than the range of long-term means for that climatic parameter.

CONTENTS

While the atmosphere is the most active of all geological agencies, it has received the least careful study from geologists. It very actively destroys its relics almost as soon as formed and gives them peculiar evanescence. This has invited the neglect of geologists laudably prone to concentrate their attention upon agencies which have left enduring and unequivocal records. The atmospheric element in geologic history bids fair to long remain obscured by elusive factors and uncertain interpretations. None the less it is an element of supreme importance and should be persistently attacked until it yields up its truths. This must be my excuse for offering a paper which, I am painfully aware, is very speculative in many of its parts.

T. C. Chamberlin, 1897

INTRODUCTION

Study of rates and forms of mass wasting in the Sierra Nevada has revealed that past variation in over-all rates and dominance of one form of denudation over another seems to be a function of climate. This paper proposes a paleoclimatic model for the last 10,000 years, based upon numerous techniques of paleoclimatic analysis. In a later paper evidence will be presented to demonstrate that the late Holocene climatic trends in the Sierra Nevada, although more complex than previously thought, are approximately synchronous with like climatic trends throughout much of the world.

For the purposes of this paper, the Pleistocene-Holocene boundary will be assumed to be 10,500 b.p.[1] This date is based upon papers and correspondence submitted to the Subcommission on the Holocene of the International Congress on the Quaternary. Zagwijn (1965, written commun.) and Nilsson (1965) both suggest that 10,300 to 10,500 be accepted as the date for this boundary, based upon glacial stratigraphy and paleoclimatic work in northwestern Europe. This is in close agreement with estimates made from the data presented by Adam (1967, p. 291, 1966, personal commun.) for the last major retreat of ice from the latest Pleistocene maximum in the Sierra Nevada. Work by McCulloch and Hopkins (1966) in northwestern Alaska and by Kauffman and McCulloch in the Rocky Mountains of Colorado (1965, p. 1228) also support this date, although it is not in agreement with the ideas of some other workers (Richmond, 1965, p. 226; Neustadt, 1967). All of the previously mentioned work that supports the *circa* 10,500 b.p. arbitrary

---

[1] Almost all ages between 3000 B.P. and 50,000 B.P. that are mentioned in this paper are directly or indirectly based upon C^{14} age estimates made on or before January 1, 1967. Since such age estimates are based upon at least two false assumptions (that the half life of $C^{14} = 5568$ years and that the ratio C^{14}/C^{12} has been constant from 50,000 years ago up to the beginning of the industrial revolution), a radiocarbon age estimate cannot be equated with a calendar date or a specific number of "years ago." In this paper all radiocarbon-based "ages" will be referred to as b.p., b.c., or a.d., whereas calendar-equivalent age estimates will be referred to as B.P., B.C., and A.D.

boundary is based directly or indirectly upon an estimation of the age of the midpoint between the maximum and minimum temperatures of the last major north-temperate-latitude glacial-interglacial oscillation. This climatic definition is in accord with the original definition of the Holocene as that period in which conditions can be closely compared with those prevailing at present (Zagwijn, 1965, written commun.). Neustadt (1967) reviews the 1961 INQUA Commission data and proposes that the beginning of the Allerød (±12,000 b.p.) be accepted as the beginning of the Holocene.

The Sierra Nevada is a large asymmetric tilted and faulted mountain block trending north-northwest for more than 600 km in eastern California. It averages 120 km in width and is bounded on the north by the southernmost extension of the Cascade Range at 40° N. latitude and on the south by the Mojave Desert at 35° N. latitude. The gentle western side of the range rises from the Sacramento–San Joaquin Valley at 100 to 200 m elevation to crest altitudes of 2500 to 4400 m near the eastern edge of the uptilted block. The eastern escarpment is a dramatic mountain wall rising 750 to 3100 m above downfaulted and downwarped valleys of the Basin and Range Province.

The Sierra Nevada was extensively glaciated during the Pleistocene (Bateman and Wahrhaftig, 1966, p. 158–169; Wahrhaftig and Birman, 1965, p. 305–310) and has recently yielded evidence of glaciation during the earliest Pleistocene, more than 2.7 m.y. ago (Curry, 1966, 1968b). That portion of the Sierra Nevada above the lowest orographic firn limit of the late Pleistocene glacial maxima is generally termed the High Sierra (Fig. 1). This orographic firn limit ranged from 2600 m on north-facing headwalls in the northern part of the study area to 3700 m on southwest-facing slopes at the southern end of the area. The generally higher climatic firn limit, defined on the basis of the elevations of the lowest late Pleistocene cirques on south-facing or neutral slopes, generally occurred at progressively higher elevations eastwardly across the range. Presumably, this was due to the effect of east-flowing air masses becoming progressively depleted in moisture as they were forced up the asymmetric mountain block.

Previous interpretations of the climatic history of the Sierra Nevada have been based upon glacial stratigraphy and variations in extent of residual cirque glaciers (Matthes, 1940, p. 398–401), on interpretation of tree-ring width variations from trees growing in areas adjacent to the Sierra Nevada on the simplified assumption that narrow rings indicate periods of drought (Fritts, 1965a), and on study of the histories of lake levels and chemistry variations for lakes along the arid eastern base of the range (Smith, 1968; Thomas, 1962, 1963). Limited palynological work is in progress based upon analysis of postglacial sediments within the High Sierra (D. P. Adam, 1966, personal commun.). Previous workers have all recognized that climatic changes accompanied transitions from glacial to interglacial climates, but none has suggested quantitative limits on what these changes may have been for the Sierra Nevada. This paper attempts to define some of these limits.

On a world-wide scale, paleoclimatic analysis of the last 10,000 years has received much attention recently. For North America and Europe, a more or less unified picture of events occurring during the last 8000 to 10,000 years has been formulated by workers using data from many scientific disciplines. At a recent conference on world climate for the period 8000 to 0 B.C., papers

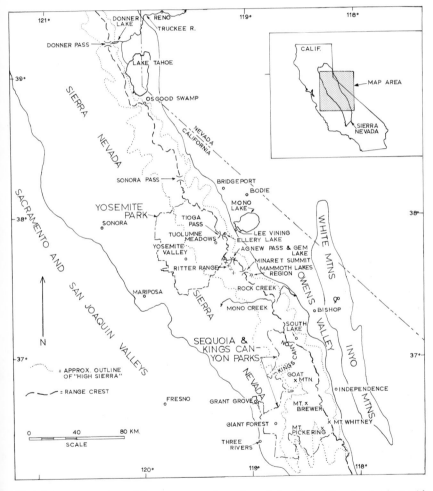

Figure 1. Location map of the central portion of the Sierra Nevada physiographic province and adjacent portions of the Basin and Range and Sacramento-San Joaquin Valley Provinces.

were presented showing, for instance, charts of mean isobars for the years 6500 B.C., 4000 B.C., and 2000 B.C. (Lamb and others, 1966, p. 194–195). Such work shows a rather consistent pattern of parallelism of major climatic trends throughout the populated areas of the northern hemisphere. Analysis of past and present climatic trends based upon actual records and interpretation of written historical records has long been popular. Williamson (1770, p. 336) discussed climatic trends in North America, and Mann (1779, p. 337–347) perceptively reviewed data on European and American trends for the last 2000 years. Comprehensive recent work on regional and world-wide climatic trends include that by Betin and Preobrazenskij (1962) and Sawyer (1966).

ACKNOWLEDGMENTS

Funds for direct support of portions of this research were provided by the Geological Society of America, the National Science Foundation, and the University of California at Berkeley. Original photographs by G. K. Gilbert, I. C. Russell, J. N. LeConte, F. P. Farquhar, F. E. Matthes, W. D. Johnson, Ansel Adams, Stephen Willard, M. R. Parsons, Oliver Kehrlein, and Eadweard Muybridge were supplied by the Sierra Club, Bancroft Historical Library, U.S. Geological Survey, National Park Service, and the photographers themselves or their heirs.

William A. Weber, of the University of Colorado Herbarium, provided lichen identifications and herbarium space for voucher specimens. Clyde Wahrhaftig, Garniss H. Curtis, Jack Major, Stephen C. Porter, and Robert P. Sharp provided distinctly helpful criticism of various manuscript drafts. Field assistance was supplied by my wife, Nancy E. Curry. This paper is a summary of a portion of a Ph.D. dissertation at the Department of Geology and Geophysics of the University of California at Berkeley.

PRESENT CLIMATE

Paleoclimatic studies are often justly criticized because too little is known of the present climate of a site to make valid comparisons with the past. As a standard for comparison with past climatic conditions, something must be known about the present means and extremes of the climate of the Sierra Nevada.

Actual Climatic Data

Published climatic data for the High Sierra include snow-survey records for a comprehensive network of stations observed since about 1930 (California Department of Water Resources, 1965) and snowfall data for Donner Pass since 1870 (Heald, 1949, p. 62). Standard temperature and precipitation records are available for three similar eastern-slope High Sierra stations. The period of record for the climatic normal (1931–1960) is summarized for one of these stations (Ellery Lake, 2900 m) on Plate 1. Additional data for other High Sierra locales are presented by Clausen (1965, p. 1318), Miller (1955), Dale (1959, p. 26–29), and the U.S. Army Corps of Engineers (1957). Plate 1 also includes data summaries for two stations on the eastern and western sides of the Sierra Nevada that illustrate the over-all regional climatic pattern for east-central California.

The High Sierra delimits the boundary between the cool, moist Pacific air masses on the west and the warm, dry continental conditions of the Basin and Range Province on the east. The subalpine and montane regions of the western Sierra thus are dominated by marine climatic influences while the alpine, subalpine, and montane regions of the eastern Sierra are dominated by continental weather. Alpine regions of the western Sierra and range crest appear to be transitional in character. This difference is reflected by the marked precipitation differences between Yosemite National Park and Bishop (Pl. 1). Both stations are at nearly the same elevation, but Bishop is in the rain shadow of the mountain range.

Use of the terms "alpine," "subalpine," and "montane" is based upon the philosophy that climax vegetation most accurately reflects climatic parameters. Climax regions used in this paper are approximately similar to the climax vegetational zones described by Oosting (1958, p. 273–312) for the Sierra Nevada. Stone (1962, p. 15) and Munz and Keck (1963, p. 10–18) give additional data on probable climatic factors typical of these vegetation units.

The entire Sierra Nevada has a very marked wet and dry season due to the migration of the semipermanent Pacific high-pressure area southward in the winter—allowing cool, moist Pacific storms to penetrate into the Sierra—and northward in the summer—deflecting Pacific air masses farther north. The proximity of the Sierra Nevada to the Pacific Ocean results in a disparity between the temperatures of the Pacific air masses and adjacent ground temperatures, the air being warmer than the ground surface in the winter but cooler in the summer. This maritime influence results in moderation of temperature extremes throughout the Sierra and also contributes to the marked summer drought.

Precipitation in the High Sierra occurs mainly as snow between October and May. Summer precipitation in the High Sierra is dominantly limited to locally heavy, convectional thunder-shower activity associated with warm, moist air that moves northward into the southern regions of the Sierra Nevada from southern California and the Gulf of California.

In winter the disintegration of the general summer continental low-pressure cell over western North America causes the Pacific high to break up or move far to the south, thus allowing Pacific air to flow eastward across the

Sacramento–San Joaquin Valley into the Sierra. Storms from the semi-permanent Aleutian low then carry precipitation into the Sierra—generally in the form of snow.

Winter precipitation increases with altitude on the windward (western) slopes of the Sierra Nevada up to a certain elevation, then decreases to the crest and eastward on the leeward (eastern) side, and desert conditions prevail along the eastern base of the range. The snow-survey data permit an approximation of the elevations of regions of maximum precipitation in the western Sierra Nevada. At Donner Pass (2200 m) snow depths increase continuously to the top of this unusually low pass. Farther south, in northern Yosemite National Park, maximum snow depths are consistently recorded at an elevation of about 2600 m (Fig. 2, A–A′), while in the southern part of the range maxima are found at 2500 m (Fig. 2, B–B′). The area of maximum water content of the snowpack generally coincides with the area of maximum snow depth. Winter snow-depth maxima usually occur in April or May throughout the range and, although highly variable, average about 2.5 m in the southern part of the range and 3.5 m in the northern part at the elevations of maximum accumulation.

Indirect Climatic Indicators

Present High Sierra climatic conditions are indirectly reflected by vegetation, position of the firn limit on residual glaciers, range of active patterned ground and permafrost, range of seasonal and diurnal ground-freezing phenomena, and bedrock temperatures. These climatic indicators have been observed during the course of 265 field days in the central portion of the High Sierra between Sonora Pass and Mt. Whitney during the years 1965 and 1966. Inferred data from many of these indirect sources have been combined with the limited direct climatologic data to produce the climatic profiles shown in Fig. 2. The data from which these interpretations are made as well as discussion of climatic implications of the distribution of permafrost and regional tree limits in the Sierra Nevada are covered by Curry (1968a, p. 12–17).

Present Glaciers

Glaciers are found in cirques facing north and east in the higher regions of the Sierra Nevada. Because they are below the present climatic firn limits, the locations of these small (less than 2.5-km^2 area) ice bodies are determined by sites of net snow accumulation rather than any simple solely temperature-dependent variable. Lowest existing orographic glacier firn limits (as defined by Flint, 1957, p. 48) during 1965 varied from 3050 m in the Ritter Range at 37°36′ N. latitude to 3300 m on the northern face of Mt. Gardner at 36°48′ N. latitude. At 37° N. latitude, however, some glaciers were found to have firn limits as high as 4300 m whereas on what is probably one of the southern-most glaciers in the Sierra Nevada, on the northern side of Mt. Pickering at

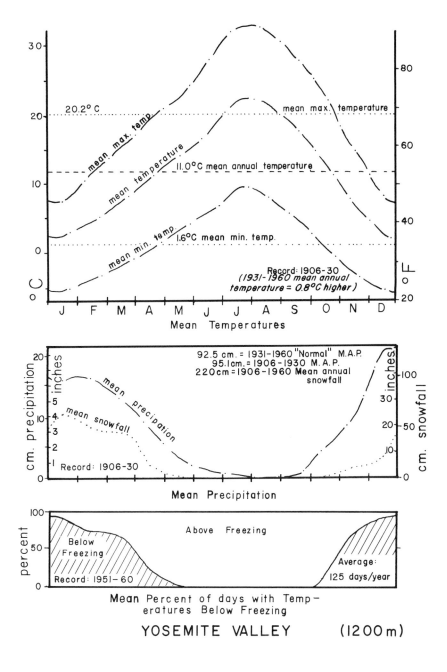

YOSEMITE VALLEY   (1200 m)

36°32′ N. latitude, the 1965 firn limit was at 3900 m. Flint (1957, p. 47) suggests that the present regional climatic firn limit for the Sierra Nevada would be found at altitudes of 4500 to 4700 m—at least 100 m higher than the summits of the highest peaks. These firn-limit data were observed during a period when precipitation and temperature means were close to those of the 1931 to 1960 climatic normal.

TECHNIQUES OF PALEOCLIMATIC STUDY

The techniques employed to investigate the Holocene climatic history of the Sierra Nevada include (1) review of the written and photographic records of the last few centuries, (2) analysis of variation in tree rings, (3) lichenometric analysis of the ages of deposits formed within cirques and along the walls of glaciated valleys, (4) study of observed vegetational age classes, (5) study of the prehistoric variation in the position of timberline, and (6) interpretation of the published data on Holocene Sierran palynology. One early Holocene deposit thought to reflect a changing climate was dated by radiocarbon.

The above techniques are used to determine periods of partial retention of winter snowpack throughout the summer, as well as periods of minor advance of small cirque glaciers. Known historical data on the climates of these periods of minor glacial advance in the High Sierra are then compared by probability analysis with the variance of actual climatic data for the entire period of historic record. This comparison helps to determine the actual climatic conditions that might be typical of a "glacial" climate during times of Holocene glacial advance.

These techniques provide a consistent picture of the climate of the High Sierra during only the last 6000 years. Generalizations of climate before about 6000 B.P. are based upon radiometrically dated glacial deposits within the Sierra and on published data on early Holocene climate in adjacent areas.

Historical Records

Routine snow-depth measurements and spot temperature and precipitation data have been recorded for various parts of the entire Sierra Nevada since approximately 1865. Most of the earliest data are in the form of discontinuous records published in the newspapers of the Sierra Nevada mining camps between 1860 and 1810. Few of these recording stations were supervised by the U.S. Weather Bureau, and many of the data have never been published in official sources. Monthly snowfall has been recorded continuously since 1870 at Donner Pass (2200 m); this constitutes the longest continuous record of this type for the United States (Heald, 1949, p. 62). Spot "thermometrical measurements" as well as qualitative statements about precipitation and temperature are recorded in the newspapers of the Mother Lode district

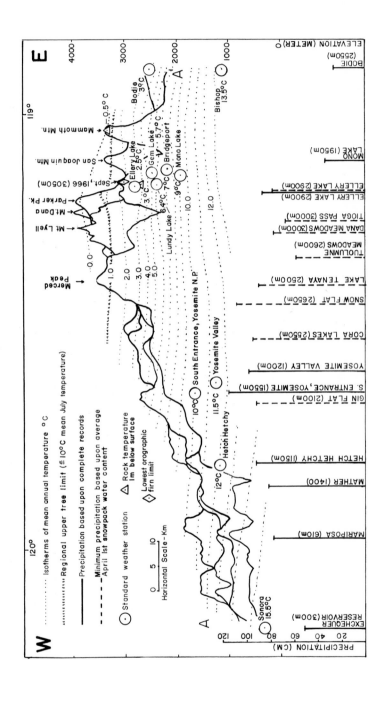

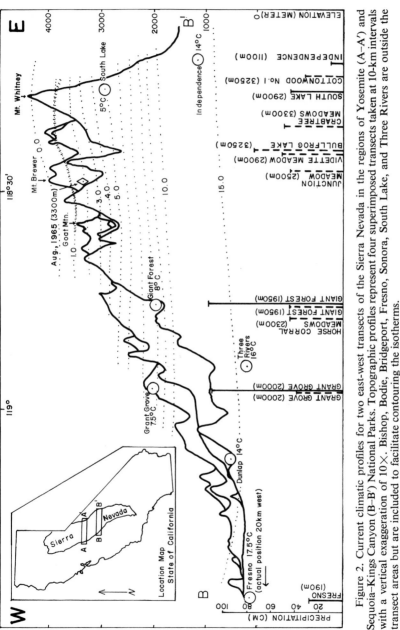

Figure 2. Current climatic profiles for two east-west transects of the Sierra Nevada in the regions of Yosemite (A–A′) and Sequoia–Kings Canyon (B–B′) National Parks. Topographic profiles represent four superimposed transects taken at 10-km intervals with a vertical exaggeration of 10×. Bishop, Bodie, Bridgeport, Fresno, Sonora, South Lake, and Three Rivers are outside the transect areas but are included to facilitate contouring the isotherms.

in the western foothills (450 to 800 m elevation) and for areas such as Bridgeport (2000 m), Bodie (2500 m), and Mammoth Lakes (2700 m) along the eastern flank of the range. These data are by no means complete or unbiased, but when coupled with more reliable long-term records begun in the period 1849 to 1880 at such sites away from the Sierra as Sacramento and Independence, as well as with snowfall records at Donner Pass, the newspaper accounts can substantiate periods of major departure from long-term means during the last 50 to 100 years.

Official U.S. Weather Bureau data for the eastern side of the Sierra, including some mining camps within the High Sierra, are summarized by the Weather Bureau for the period from the establishment of the stations to 1930 (U.S. Weather Bureau, 1932). Temperature and precipitation data for the High Sierra are actually more complete for the period 1880 to 1910 than for any later time due to the widespread mining activity during that time. Historical records of the climate of the Sierra Nevada before 1870 consist merely of the accounts and written journals of early travelers in the mountains.

Although a period of drought in San Francisco and Sacramento cannot necessarily be equated with similar conditions in the Sierra, journals of the early mountain explorers that suggest unusually dry winter seasons can be coupled with the long-term records from the population centers to arrive at generalizations back to 1849 (Dale, 1959) or coupled with the earliest historical data on lake levels east of the Sierra to extend the precipitation record back to 1839 (Hardman and Venstrom, 1941). A precipitation curve has been constructed for Los Angeles, California, back to 1769 on the basis of diaries and records of the Spanish missionaries (Thomas, 1962, p. 37).

Information on cloud cover and relative size and extent of summer snowbanks in past years can be obtained by reoccupying the sites from which the old photographs were taken on or near the same calendar date and comparing the new and old photographs. If snow conditions, lake levels, and seasonal variability of the vegetation are obviously different at the site when it is reoccupied, one can try to ascertain when, during the given season of observation, conditions would most closely approximate those of the original photography. By such means it was learned that spring and early summer of 1966 were warmer, drier, and more cloud-free than approximately 60 percent of the spring and early-summer periods between 1890 and 1930, but the summers of 1964 and 1965 were about average. During 1966, spring and early-summer conditions were 3 to 6 weeks "ahead" of those of the pre-1930 average period of record.

In practice, since the climatic "normal" as used by the U.S. Weather Bureau is defined as the mean of those conditions that existed between the years 1931 to 1960, only photographs taken in 1930 or earlier were used for analysis of historic climatic variation. Sources of photographic data and reference to the actual photographs and other historical sources used to determine the seasonal climatic conditions that existed in the High Sierra during the late 1800's and early 1900's are listed in appendix 1 of the Ph.D. dissertation from which this paper is derived (Curry, 1968a, p. 77–96).

Figure 3 summarizes some of the better documented historical climatic data pertinent to the Sierra Nevada. It should be noted that rainfall data do not correlate well with the High Sierra snowfall. Even runoff for the Truckee River (Fig. 3, curve D), with its headwaters at Donner Pass, does not exactly parallel total snowfall at Donner Pass since warm rains may result in high runoff with only average snowfall and moderate runoff may occur during years of low snowfall due to temporary water storage in the large lakes in the

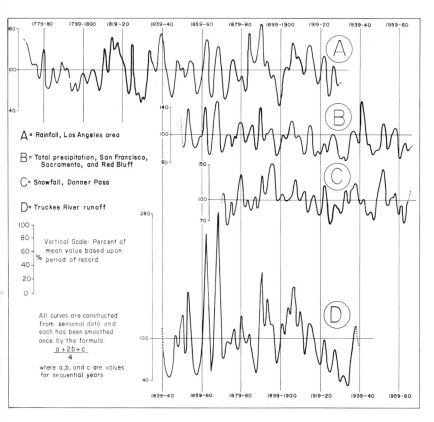

Figure 3. Summary of seasonal climatic data pertinent to the Sierra Nevada. Data for curve A are from Lynch (1931, p. 31). Curve B has been constructed by averaging deviations from the means for San Francisco and Sacramento since 1849 and for Red Bluff since 1877. Curve C is from Heald (1949, p. 62) and U.S. Weather Bureau reports. Curve D is constructed from data presented by Hardman and Venstrom (1941, p. 85–87).

headwaters of the Truckee River. Total seasonal snowfall seems to be a function of both temperature and precipitation. Since the majority of the precipitation that falls in California is derived from cyclonic storms that originate in the North Pacific, one can assume that most of the large storms that reach Los Angeles will have resulted in heavy Sierran precipitation. Times of high precipitation in northern California thus correlate with times of above-average Los Angeles precipitation. Other conditions being equal, the amount of precipitation from North Pacific storms that falls at a site in California is approximately proportional to the distance to that site from the northwestern corner of the state. Exceptions result from storms that sweep into the southern California mountains and deserts from Mexican waters, but these storms average less than two a year.

Dendroclimatology

The variation in tree-ring widths can be interpreted to extend knowledge of climatic variations back to more than 1000 years B.P. Tree rings are presently receiving a great deal of study as climatic indicators, especially at the University of Arizona Laboratory of Tree Ring Research under the leadership of H. C. Fritts. Other authors (Bray and Struik, 1963; Bray, 1965, 1966; Adamenko, 1963; Schove, 1954) have been working on tree rings in subarctic and subalpine environments. Basically, two major variables are thought to affect relative annual tree-ring width. Fritts and his temperate-latitude colleagues (Fritts, 1965a, 1965b, 1966) interpret tree-ring variations as a function of precipitation and the ratio of precipitation to evaporation as did their predecessors working with data from the western United States (Schulman, 1956; Douglass, 1936). Bray (1965, p. 440) interprets subarctic trees as "organic thermographs" rather than precipitation indicators because he assumes that moisture is seldom a limiting factor in tree growth during the subarctic growing season. Both philosophies are no doubt partly correct within their environments, but, as was recognized in 1925 by Antevs (p. 117) and Huntington (1925, p. 157) working with the giant sequoias of the western Sierra Nevada, many other factors could affect tree-ring widths.

For this study, cores from 49 High Sierra trees growing in diverse sites were analyzed for paleoclimatic data. Only those trees found to respond consistently to established climatic variables in the last 100 years were used to extend the record of climatic change beyond the period of historical record. Representative cores are plotted in Figure 4. These were selected to illustrate the complete range of variation patterns noted. Multiple cores are analyzed from each tree to lessen the effect of double and missing rings.

Five arboreal species in a variety of environments were investigated to determine the relations between ring width, climatic variables, tree species, and site. By comparing a tree growing at the edge of a permanent lake or

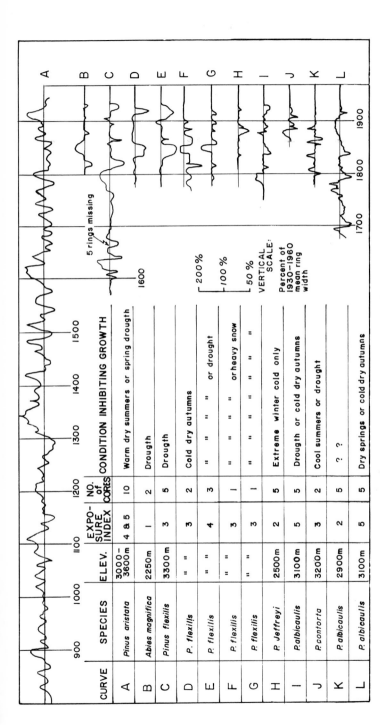

Figure 4. Representative tree-ring growth-rate curves for the Sierra Nevada (curves B–L) and the White Mountains (curve A). All Sierran data are from the High Sierra except curves B and H. Sierran curves represent annual ring widths plotted as a function of mean ring width for the period 1930 to 1960. The first 20 years of a tree's growth are not shown. Curve A is from Fritts (1965a, p. 428–432).

CURVE	SPECIES	ELEV.	EXPO-SURE INDEX	NO. of CORES	CONDITION INHIBITING GROWTH
A	Pinus aristata	3000–3600m	4 & 5	10	Warm dry summers or spring drought
B	Abies magnifica	2250m	1	2	Drought
C	Pinus flexilis	3300m	3	5	Drought
D	P. flexilis	" "	3	2	Cold dry autumns
E	P. flexilis	" "	4	3	" " " " or drought
F	P. flexilis	" "	3	1	" " " " or heavy snow
G	P. flexilis	" "	3	1	" " " " " "
H	P. Jeffreyi	2500m	2	5	Extreme winter cold only
I	P. albicaulis	3100m	5	5	Drought or cold dry autumns
J	P. contorta	3200m	3	2	Cool summers or drought
K	P. albicaulis	2900m	2	5	? ?
L	P. albicaulis	3100m	5	5	Dry springs or cold dry autumns

5 rings missing

VERTICAL SCALE:
Percent of 1930–1960 mean ring width

200%
100%
50%

stream with one of the same species growing in a nearby area subject to both standard low-water-table drought and "physiologic" drought, one can theoretically eliminate the variability of drought and note the effects of other variables on the growth of the trees. Physiologic drought is a desiccation of tissues by cold fall or winter winds that occurs when early freezing of the ground prevents the physiologic adaptation of the tree for very low temperatures. This destruction of conifer needle tissue by cold winds is usually reflected in tree rings by inhibited growth for several years following the incident.

Pinus flexilis James and P. albicaulis Engelm. growing on the crests of ridges faced with free, well-fractured, permeable cliffs have the least chance of access to a permanent summer water source, yet these species also tolerate conditions where their roots are continuously below a water table. These species thus afford the best possibilities for recording the recent paleoclimatic history of the High Sierra.

Correlation of variation in tree-ring width with a simple model of past variation in moisture and temperature proved very difficult, even for trees of a given species growing at a single elevation within one valley. Clearly, factors such as insect infestations, depth of winter snow, and winter frost damage must affect ring widths and ring frequency (missing and double rings) in the high mountains, in addition to temperature and precipitation. The variable species-specific factors upon which the interpretation of the Sierran tree-ring variations is based are reported elsewhere (Curry, 1968a, p. 23–26).

Fritts (1965a) presents analyses of tree rings of Pinus jeffreyi Grev. & Balf. in A. Murr. from the eastern base of the Sierra Nevada and from P. aristata Englem. from the timberline region of the White Mountains just east of the Sierra Nevada. The White Mountain data have been plotted at the top of Figure 4. These data, collected by Schulman and published by Fritts, are presented as deviations of 10-year means from the mean for the period 1651 to 1920 A.D., plotted at 5-year overlapping intervals. Thus, for this particular curve, long-term changes will be attenuated and short-term fluctuations will be emphasized. The White Mountain data are especially valuable since they are the result of so much careful work and record such a long interval of time.

The use of tree rings for paleoclimatic analysis in diverse mountainous environments requires the consideration of so many variables that the results must certainly be subject to interpretive errors. Nevertheless, analysis such as that used in this report has proven of value when used in combination with other paleoclimatic indicators.

Tree-ring analysis alone suggests that periods of above-normal precipitation have occurred several times since the beginning of the record (which begins at about 800 A.D. for the White Mountains and 1600 A.D. for the

High Sierra). The period 800 A.D. to 1300 A.D. appears to have been one of very little climatic fluctuation and probable general drought except for brief wet periods at about 900 A.D. and 1100 A.D. Beginning in 1300, the climate began to oscillate rather widely with frequent periods of markedly above-average precipitation, especially during the first half of 1300, in the 1440's, the 1470's, the 1530's, and the first half of the 1600's. More recently, wet periods have occurred in the 1860's, the 1880's, and the 1890's. Since both extreme wetness and dryness seem typical of over-all wetter periods, the low amplitude of the oscillations between 900 and 1050 A.D. and between 1150 and 1250 A.D. could imply periods of drought then. Marked drought can be interpreted as having occurred during the periods 1870 to 1880, 1840 to 1860, 1770 to 1810, and possibly in the early 1700's.

Lichenometric Analysis

Lichenometry has been used to establish a chronology of events that are believed to be related to climatic variations. Periods of glacial advance, increased rates of mass wasting, and periglacial activity involving formation of sorted and nonsorted frost polygons and solifluction lobes have been dated by study of the lichen cover on deposits dating from these periods or on bedrock overridden by glacial ice.

The use of lichen cover to date the last period of disturbance of a rock surface has been discussed by Beschel (1961), Benedict (1967, 1968), and Andrews and Webber (1964). The technique involves (1) the establishment of a growth-rate curve for one or more species of lichens, and (2) the measurement of the diameters of the largest individual lichens of those species on the rock surface of unknown age, on the assumption that the size of the lichen is directly related to its age by the growth-rate curve. Rock surfaces up to 6000 years old may possibly be dated using extremely slow-growing species.

Because lichens would not be expected to survive transport through a glacier, there is little chance of an anomalously old lichen age for a true glacial moraine. For rock glaciers where a lichen-covered boulder on the cirque headwall may be carried down the glacier without burial for more than three seasons, however, it is sometimes necessary to construct a lichen size-frequency curve and consider that the age of the material at a given point on a rock glacier is equivalent to the age of the lichens that comprise the oldest mode on the frequency curve. Benedict (1967, Figs. 8, 9) graphically illustrates how size-frequency analysis is used to detect disturbed or inherited lichen cover. Sampling techniques used in the present study are similar to those of Benedict (1967, p. 827–828).

In the Sierra Nevada the following lichen species were found to be useful for lichenometric analyses: *Acarospora chlorophana* (Wahlenb.) Mass., *Rhizocarpon superficiale* (Schaer.) Vainio, *R. lecanorium* Anders., *Sporastatia testudinea* (Ach.) Mass., *Lecidea atrobrunnea* (Ram.) Schaer., and *Lecanora muralis* (Schaer.) Rabenh. Lichen species were identified by W. A. Weber at the University of Colorado and voucher specimens are on file in the herbarium of that university.

Acarospora and the two species of *Rhizocarpon* were found to reach maximum size in high alpine areas only on surfaces that had not been disturbed by any of the generally recognized neoglacial ice advances. These species were thus singled out as potentially the longest lived, slowest growing individuals for which detailed growth curves should be established. The other aforementioned species were less valuable but were used for dating surfaces less than about 2000 years old and to help establish the growth-rate curve for the slower growing species by ratio techniques (Andrews and Webber, 1964; Løken and Andrews, 1966). All species used are crustose or squamulose, that is, they grow firmly attached to the rock substrate and are seldom more than 2 to 3 mm thick.

Little is known about the autecology of lichens, that is, how they react to their environment and how, for instance, their growth rates may depend upon microclimatic variables. Growth depends upon the amount of time that humidity, light, and heat exist in adequate proportions for a given species in a given stage of its development (*see* Lange, 1965, p. 1). Comparative studies of some rapidly growing lichen species have suggested that their rate of growth is not linear and may vary with age (Hale, 1961, p. 40). Beschel (1961, p. 1055) has introduced the concept of "hygrocontinentality" or variation in growth rate as a function of climatic variables roughly equivalent to continentality to explain variations in growth rates that he has noted for lichens of a given species growing in northeastern Canada, Greenland, and the Alps. My own work and that of Benedict suggest that lichen growth rates are at least in part a function of the length of the lichen's growing season—which in many cases does not vary as a function of continentality. The growing season is a function of the length of the snow-free period and the possibility of microclimatic drought during that snow-free season. This has also been recognized by Beschel in his recent work on arctic lichen growth-rate variations (Beschel, 1965, p. 26).

Generally, both the length of the snow-free season and the possibility of microclimatic drought decrease with increasing altitude. Since these two factors affect lichen growth rates in opposite manners, growth rates in the Sierra Nevada may remain rather constant over the range 2500 to 4000 m elevation for a given topographic exposure. Marked exceptions are to be found on windswept mountain summits where discontinuous growth may be possible throughout the winter and along the banks and beds of streams or rivulets where lichens may remain continuously moist for a major portion of the snow-free period.

Comparison of the maximum diameters attained by various lichen species on moraines of a given age but made up of rocks of varying lithologies has shown that the species used in this study are relatively insensitive to the chemistry of their substrate. Rocks with readily soluble calcium, such as marble and some calc-silicate hornfels, give rise to exceptions and were avoided for lichen dating. Anomalous growth rates were also found for *Acarospora* growing on porous andesitic ignimbrites due to either the chemical or the physical properties of this rock and this combination of lichen and substrate was likewise avoided for lichen dating.

Growth-rate curves have been established for *Acarospora* and *Rhizocarpon* for the High Sierra by measuring the maximum lichen diameters on surfaces of known age and also by cross-dating the ages of these slower growing species with faster growing species such as *Lecidea atrobrunnea*. In addition, growth rates have been determined directly by comparing photographs of lichen-covered boulders taken at intervals of 50 to 80 years. Where lichens are clearly visible on foreground boulders in old photographs, one merely has to measure the increase in diameter of those lichens to determine their recent growth rates. Such measurements must usually be made at the photograph sites. With photograph scales of 1:1 or larger, changes in lichen diameter can be measured to ±2-mm accuracy.

Comparative studies of photographs showed that very young lichens grow faster than larger individuals for all the species investigated. Analysis of more than 500 photographs taken before 1910 in the High Sierra above 2500 m elevation showed that the first small lichens will not become established on a fresh granitic surface for between 20 and 50 years. This is so, generally, only where wind must act as the agent of dissemination of the colonizing crustose lichens. Along stream banks, lichens may appear within 10 years. This colonization period must be added to the age of the oldest lichens on a deposit to determine the age of that deposit.

Surfaces of known age in the highest portions of the mountains were generally only 100 years old or less and were associated with mines, mineral prospects, and road building. Such surfaces were dated by historical references in newspapers and county mining records. Moraines formed or overrun by ice in 1895 (dated from photographs) were also used to establish growth rates during colonization by pioneer species. Another fixed point was estimated by dating, by dendrochronometry, numbers of glacial moraines whose vegetational cover seems to have been established about 950 B.P. Although these moraines may not have formed at that time, the age of trees growing upon them suggests that the present vegetational cover probably dates from that period. By assuming that 50 to 100 years are required for the establishment of the first trees on a lichen-covered High Sierra moraine, the age of the lichens was estimated as 1000 to 1050 years. Numerous control points in the range of

100 to 500 years have been gained by study of the lichen cover on split boulders and exfoliation slabs where the age of a portion of the rock surface may be estimated by counting the rings in tree roots. To be reasonably certain that the root was responsible for the removal of the rock slab upon which lichens became established, one must show that the presently exposed portion of the root was once confined to a narrow space where lichens could not be expected to grow. This can usually be ascertained by sectioning the root and noting the shape of the oldest central rings relative to those that formed after the overlying slab was forced off. Another method of dating lichen cover, helpful in corroborating an established growth-rate curve, involves dating the occurrence of a destructive forest fire by tree-ring analysis of surviving trees and assuming that the lichen cover on fire-spalled rock slabs was established since the time of the last such major fire.

   Lecidea atrobrunnea and *Lecanora muralis,* two species that grow more rapidly than *Acarospora* or *Rhizocarpon,* were found by photograph comparison to increase in diameter by 33 and 30 mm per century, respectively. This growth rate was established in the timberline or higher regions of the Sierra and was linear for individuals between about 100 and 1200 years of age. Since *Lecidea* and *Lecanora* grow rapidly enough to establish quite accurate growth curves during the 100-year period when accurate photograph comparison is possible, these species themselves can then be used to date older surfaces up to 1200 years old so that more points may be added to the *Rhizocarpon* and *Acarospora* growth-rate curves.

   Although no moraines less than 5000 to 6000 years old were cross-dated by both carbon 14 and lichenometry in the Sierra Nevada, some rhyolite domes at 2800 m elevation east of the main Sierra crest have been dated by potassium-argon and thorium 230 and are within the potential dating range of the slowest growing lichen species on the soundest bedrock (Tadeucci and Broecker, 1969; Dalrymple, 1969). Although both the radiometric and lichenometric dating techniques may be giving dates that are as little as one-half the true age of the rhyolite cooling surfaces, general agreement in the range of 5000 to 6000 years ± 1000 years suggests that the noted rate of growth of a 500-year-old *Acarospora* may remain relatively constant for a long period under ideal conditions.

   Figure 5 is a growth-rate curve that was found to fit most closely all the data for *Acarospora chlorophana* and the two species of *Rhizocarpon* studied. That all three species should grow at about the same rate is remarkable, but the differences were not statistically significant, and this single curve may be considered as applicable to all three species subject to the restrictions of rock type and site characteristics mentioned earlier. Published growth curves for alpine crustose lichens have been based upon the assumption that the growth rate is essentially constant with respect to time within the accuracy of the control dates available. Where sufficient historically dated surfaces are available, initial growth has been found to be more rapid, so over-all growth-rate curves are logarithmic (*compare* Benedict, 1967, p. 830). The reverse situation, where rate of growth increases directly as a function of the size of the lichen, has been suspected (Hale, 1961, p. 40) but not noted in any species studied here.

The growth curve presented in Figure 5 is unlike other published lichen growth-rate curves in that growth is represented as a series of sigmoidal steps instead of the usual linear rate. Control data are not such that the actual shape of this curve is restricted to that form presented; however, this curve is in agreement with the observed variation in growth rates noted for the last 50 years as a function of variations in site moisture. The positions of the inflection points of more rapid growth are determined by the average lichenometrically determined ages of glacial advances, while the slope of the curve is estimated

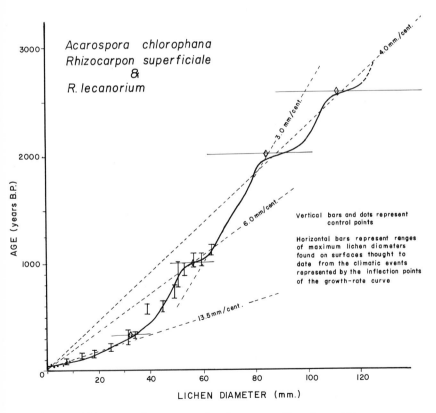

Figure 5. Lichen growth-rate curve for *Acarospora chlorophana*, *Rhizocarpon superficiale*, and *R. lecanorium* for alpine and subalpine regions of the High Sierra.

on the basis of the assumption that those times of advance represent periods of moister over-all climatic conditions when growth rates would approach those observed today at moister than average sites. Thus a rough climatic model is used to adjust the shape of the lichen growth curve, but even this curve is, at best, very much simplified.

Over-all growth rates for the first 2600 years for *Acarospora* and the two *Rhizocarpon* species in the High Sierra were found to be 3.8 to 4.5 mm per century. This compares with estimates for similar species of *Rhizocarpon* of 3.6 mm per century for the alpine regions of the southern Rocky Mountains of Colorado (Benedict, 1967), 3.5 mm per century for the interior of Greenland (Beschel, 1961, p. 1048), 10 mm per century for the eastern Alps (Heuberger, 1966, p. 115), about 6 mm per century for interior Baffin Island (Andrews and Webber, 1964, p. 95), and more than 30 mm per century for southern coastal Greenland (Beschel, 1961, p. 1048). In general, the published rates above 8 to 10 mm per century are based upon control points for the last 200 to 300 years only and are not, in my opinion, necessarily representative of long-term growth.

Ages of lichens on talus, cirque headwalls, moraines, rock glaciers, solifluction deposits, and similar products of Sierran climates cooler and moister than those of the last 50 years reveal three major periods of Sierran neoglacial activity during the last 5000 to 6000 years. The accuracy of the following dates decreases with increasing age as suggested by the ranges of lichen sizes of different ages shown on Figure 5. Generally, dates within the last 1000 years are ±50 to 100 years while those between 1000 and 3000 B.P. are ±100 to 300 years.

The latest period is represented by moraines and rock glaciers dated between 1250 A.D. and the present. Moraines formed by three or more advances between 1750 and 1898 A.D. generally constitute the terminal moraines of this most recent period of neoglaciation, but talus and occasional lateral moraines bear a lichen cover that indicates that a series of three advances occurred between 1550 and 1700 A.D. Wetter, cooler conditions before 1550 A.D. are revealed only by periods of talus and scree accumulation.

No deposits attributable to periods of increased winter snowfall and cool summers were dated from the period 1050 to 1250 A.D. Several forested moraines in the High Sierra bear a lichen cover that, together with the trees, suggests that the moraines were formed between 850 and 1050 A.D. Lichen ages of 900 to 1100 years B.P. were also noted on some shallow cirque headwalls and glaciated surfaces, but such dates were not found on any talus or solifluction deposits. This suggests that a very brief period of glacial advance may have occurred about 850 to 1050 A.D. This climatic period was of sufficient severity to either kill the lichen cover on moraines formed at an earlier time or form moraines twice as far from the cirque heads as those of the glaciers of the 18th and 19th centuries. These moraines could not be dated by radiocarbon, so either explanation is possible. Whatever the age of formation of these rather rare moraines, the marked grouping of lichen dates between 850 and 1050 A.D. strongly suggests that this was a period of glacial advance or at least of snowfall of sufficient depth to destroy the lichen cover on some moraines and cirque headwalls. Such destruction would result from any period of time when a deep snowpack remained on the moraines and cirque walls for as little as three successive summers.

No moraines or mass-wasting deposits were found to have formed solely during the period 0 to 850 A.D. If minor glacial advances did occur during this period, they were overrun or their lichen covers were destroyed by the more recent advances. Detailed photographic analysis of the glacial advance of the 1890's showed that the mere presence of ice within 1 m of a pre-existing moraine killed the lichen cover on that moraine. Thus the lichen date of a moraine is actually an estimation of the date that the ice last advanced to that moraine and not necessarily the age of the bulk of the deposit.

Distinct moraines, formed during at least two periods of multiple glacial advances, bear lichen covers that indicate that they are at least 2000 years old. These frequently encountered, distinct moraines are generally each composed of two to three composite morainal ridges. They reflect repeated glacial advances out 2 to 4 times as far from the cirque headwalls as those of the 18th and 19th centuries. Where the moraine crests are within a few hundred meters of each other, all the moraines of this series of advances bear a lichen cover of about 2000 years of age. In some valleys where the moraines of this series of advances are separated by more than 300 m, the outermost moraine complex bears a lichen cover of between 2600 and 2700 years of age while lichens on the innermost moraine give the 2000-year-old date. These two lichen dates are also commonly noted on different portions of a single talus slope and suggest that at least two distinct periods of glacial climate existed between 2000 and 2700 B.P. As can be seen from the range of maximum lichen sizes found on moraines of this age (Fig. 5), greater refinement of the climatic sequence between 2000 and 2700 B.P. is not yet possible by lichenometry.

No climatically significant mass-wasting or glacial features could be found that would have formed between 2700 B.P. and 5000 to 6000 B.P., which is the limit of the possible ages of the oldest individual lichens found.

Vegetational Age Classes

One hundred randomly selected trees in each of four stands were cored in an attempt to analyze the possibility of using data on age classes to interpret past climatic periods favorable to the establishment of tree seedlings. Two stands containing *Pinus contorta* Dougl. ex Loud., *Tsuga mertensiana* (Bong.) Carr., *Abies magnifica* A. Murr., and various large shrubby willows (*Salix* spp.) were so studied in Yosemite National Park. One was located between Lembert Dome and Soda Springs at 2625 m in Tuolumne Meadows and the other on the southern flank of nearby Medlicott Dome at 2900 m. The other two stands included one of great arboreal floristic diversity 2 km east of the summit of Two Teats Mountain, at 2800 m on the northern fork of Deadman Creek, which contained *Pinus jeffreyi, P. contorta, P. albicaulis, P. flexilis, P. balfouriana*, Grev. & Balf. in A. Murr., *Tsuga mertensiana, Abies magnifica*, and *A. concolor* (Gord. & Glend.) Lindl., and another on the pumice flats in the same drainage basin at 2500 m in which *Abies magnifica, A. concolor, Pinus jeffreyi, P. contorta*, and *Populus tremuloides* Michx. were cored. Neither stand had been apparently disturbed by man and neither showed evidence of fire or major insect damage.

The two Yosemite National Park stands showed marked age stratification that, on the basis of evidence from old photographs and the nature of the growth rings of the possible temporary edaphic dominant *Pinus contorta*, I must interpret as the result of periodic destruction of the *P. contorta* by attacks of the Lodgepole needle miner (*Recurvaria milleri* Busck) and the pine bark beetle (*Denroctonus monticolae* Hopk.). The magnitudes of these insect-caused fluctuations in species dominance were so great that climatically controlled variations could not be detected.

Age stratification in the two Deadman Creek stands, at lower elevations along the eastern side of the Sierra, suggested in a crude way several periods of greater moisture when drought-intolerant species such as *Abies magnifica* could be established. These species became established in large numbers between 1430 and 1650 A.D. and again in the 1780's, 1810 to 1820, 1860's, and 1880 to 1895. These same periods were favorable for the establishment of virtually all the species cored, but, for instance, during the 1640's *Abies* seedlings could apparently survive long enough in marginal sites in the porous 1- to 2-m pumice mantle to allow the development of a root system sufficient to carry the trees through possible ensuing less favorable times. In the same site today, only the more drought-tolerant *Pinus jeffreyi* seedlings would survive long enough to get roots down through the pumice cover. The periods 1900 to 1910 and 1960 to 1965 appear to have been slightly cooler and moister than normal on the basis of *Abies* age classes; however, it is possible that the present crop of seedlings may not reach maturity if a drought occurs in the near future.

This method of paleoclimatic analysis holds some promise for one willing to core a very great number of trees and to program the raw data thus gained for statistical analysis. All of the unburned, unlogged stands of insect-tolerant species remaining in the High Sierra and vicinity should be so analyzed. Especially revealing might be radiocarbon analyses of core-wood samples from *Juniperus occidentalis* Hook. in the Tioga Pass, Silver Lake, and Kings

Canyon areas where extremely old individuals (3000 to 6000 years?) appear to be surviving at elevations much higher than that at which the species is able to reproduce today.

Timberline Position

The altitude of the timberline ecotone theoretically varies with climate. No work on past variation in tree line has been published for the High Sierra to my knowledge; however, a paper on this subject for the immediately adjacent White Mountain Range (La Marche and Mooney, 1967) documents a fossil tree line up to 150 m above the modern timberline ecotone that was formed by trees living 5800 to 3500 years radiocarbon b.p. This estimate is based upon radiocarbon dates of the outermost wood from four greatly eroded tree remnants (U.C.L.A.—1070 A–D, F, and G), combined with the number of annual rings noted in each remnant.

La Marche and Mooney interpret this older higher tree line as indicative of warmer conditions. The present (1953–1963) mean July temperature at tree line in the area of the White Mountain study (about 3350 m) can be estimated to be approximately 9.5°C (Pace, 1963). This is very close to the 10°C mean isotherm for the warmest month accepted by La Marche and Mooney from the literature (Daubenmire, 1954; Wardle, 1965) as being closely associated with the altitude of the alpine timberline ecotone throughout the world. Times of warmer summer temperatures may not necessarily coincide with times of drier winters.

In the High Sierra, similar fossil tree lines have been noted, but none have been dated by radiocarbon. Along the crest of the Sierra, between Minaret Summit and Agnew Pass, the present windswept timberline ecotone occurs at 3100 m, 60 m below a group of unburned, highly snow-blasted remnants of a higher tree line. Similar evidences of fossil timberlines are found on many of the more exposed, dry, upland summit ridges and flats near the eastern crest of the Sierra throughout the full 250-km north-south extent of the High Sierra study area.

Although the difference in tree-line elevations is not as great in the Sierra as in the White Mountains, the degree of preservation of the remnants and the general similarity of the sites where such remnants are preserved suggests that a contemporaneous downward shift in timberline has occurred in the two adjacent mountain ranges since 3500 radiocarbon years b.p.

Palynology

Pollen analysis, coupled with careful ecologic interpretation of the site of pollen deposition, offers a potentially promising tool for paleoclimatic studies of the Holocene of the High Sierra. Little careful postglacial pollen work has been published for the High Sierra to date, and none was undertaken for this study. D. P. Adam, of the University of Arizona Geochronology Laboratories, is currently working on pollen variation in the Sierra Nevada. His cores include one within the study area of this report (at Tuolumne Meadows, Yosemite National Park, 2600 m), two in the Yosemite area at lower elevations (Hodgdon Ranch, 1400 m, and Crane Flat, 1850 m), and one intensively studied site in the Lake Tahoe area of the northern Sierra (Osgood Swamp, 2000 m).

Adam's work is a most valuable aid in interpreting the earliest portion of the Holocene history. His radiocarbon date near the base of the Osgood Swamp section (9990 ± 800 b.p., 1-sigma error; Adam, 1967) establishes a limit on the youngest possible age for the last withdrawal of glaciers at lower elevations in the area of the swamp. The sediments immediately above this basal section contain pollen from plant species that one would expect in climates with warmer, although not necessarily drier, summers than those of today.

Radiocarbon Dates

Beyond the range of lichenometry and dendrochronometry, one must rely upon radiocarbon dates to establish the ages of climatic events. The radiocarbon date by Adam (1967) establishes a minimum radiocarbon age for the last withdrawal of montane elevation ice at about 9000 to 11,000 years b.p., while the White Mountain dates of La Marche and Mooney suggest a period of above-normal temperatures between 5800 and 3500 years b.p.

Near Emerald Lake, in the Coldwater Creek drainage of the Mammoth Lakes region of the Sierra Nevada, wood was recovered from till of early Holocene age. The complete extent of the till is not known, but its position in the valley sequence and its relation to moraines and tills of probable earliest Holocene age suggest that it formed during a time of glacial advance of only slightly greater magnitude than those of the 2000- to 2600-year-old advances. The till possesses a weakly developed azonal soil similar to that found on undated early Holocene moraines in the region. It is possible that this dated till correlates with the undated Hilgard glaciation of Birman (1964, p. 46), but, on the basis of my own Sierran Holocene stratigraphy (Curry, 1968a, p. 29x), I regard this till as younger and distinct from the Hilgard.

The recovered wood, a portion of a crushed root of one of the pine species found in timberline sites today, yielded a radiocarbon age of 7030 ± 130 years b.p. (1-sigma error, I-2287). It was recovered from 64 cm beneath the ground surface at an elevation of 2950 m in what is possibly a ground moraine 1.3 km from the cirque headwall source of the till. The carbonized wood lay 33 cm beneath the surface of the till, which was oxidized to a depth of about 20 cm. Overlying the till are 31 cm of recent volcanic ash and pumice. Within 15 m of this site the oxidized surface of the till is overlain by an extensive landslide or rock-glacier deposit made up of boulders bearing a lichen cover indicating an age of greater than 2600 years. Deposits representing the three lichenometrically dated later neoglacial periods of advance are found to overlie the landslide or rock-glacier deposit as one progresses up the valley toward the cirque head. Thus, the dated till must represent a glacial advance which overrode trees that grew near the cirque headwall about 7030 years b.p. Tree line in the local region today is at about 3200 m, so climatic conditions immediately prior to the advance that buried the wood must have been at least as warm as those in the region today. Since the study of paleotimberline in the White Mountains suggests that warmer conditions existed by 5800 b.p., the glacial advance represented by the till in the Coldwater Creek drainage must have occurred between about 6000 and 7000 years b.p.

Multiple small glacial advances during the period from 6000 to 7500 years b.p. are well substantiated by radiocarbon in the Alps (Heuberger, 1966; Mayr, 1964, 1968; Patzelt, in prep.). The radiocarbon dates in this range may be in error by as much as 600 to 1000 years (Ferguson and others, 1966, p. 1173). Thus, the advance or advances between 6000 and 7000 years b.p. may have actually occurred at any time between 6500 and 8000 years B.P.

At least one additional early Holocene period of glacial advance is recognized in the Sierra. This advance was recognized by Birman (1964, p. 46), who considered it less than 4000 years old. I consider that this period of advance, which was termed the Hilgard glaciation by Birman, occurred either between 12,000 and 13,000 years ago or about 9000 years ago on the basis of Sierran pollen stratigraphy, weathering characteristics of the till, and analogy to its radiometrically dated morphological counterparts in the European alpine sequence (Heuberger, 1966, and 1967, personal commun.). (See Curry, 1968a, p. 27x–30x for further discussion of the bases of this consideration.) The moraines are composed of masses of chemically degraded gruss and small blocks and are distinctly finer grained than either late Pleistocene tills or those formed since 7000 years b.p. The finer grained character of the moraines suggests that the glacial advance followed a period of intense chemical weathering without intensive frost action. The climatic optima of 3000 to 6000 years b.p. and 7500 to 9000 years b.p., which were periods with climate little different from that of the 1960's except for slightly warmer summer temperatures, would not be times of particularly intense chemical weathering. Dominantly chemical weathering would be favored in a warm, moist summer climate such as that suggested by the flora associated with Adam's pollen cores in the general range of 9000 to 10,000 years b.p. During this period, Sierran montane and subalpine floras consisted of warmth-loving, drought-intolerant, cold-tolerant species similar ecologically to the European floras during and immediately following the Allerød interval 11,000 to 12,000 years ago and during the earlier Bølling interstadial.

Diligent search has not uncovered datable materials from the Hilgard moraines or their possible European equivalents. This may be due to the fact that Hilgard advances overrode brushy heaths and marshlands rather than forests as did the advances of about 6000 to 7000 years ago. Thus the dates for this earliest Holocene (or latest Pleistocene) fluctuation are conjectural at present.

Relation between Snowfall and Historic Glacial Advances

Comparison of seasonal snowfall records for the Sierra Nevada since 1870 and photographs of the Sierran glaciers during the same period shows that general glacier advance and retreat appears to directly correlate with snowfall. As illustrated in Figure 3, seasonal California rainfall indices, seasonal runoff of the Truckee River, and mean annual state-wide temperature trends do not directly correlate with seasonal snowfall, which is defined as the total of daily depths of snowfall from July through June. Snowfall is not a simple function of depth of snowpack, general precipitation, or water content of snow. High snowfall occurs only during seasons of both above-normal winter mountain precipitation and near-normal to slightly above normal mountain temperatures. During periods of unusually high precipitation in central California, such as 1940 to 1941 when precipitation was 149 percent of long-term normal, snowfall at Donner Pass was actually below normal.

Photographs and surveys of the High Sierra glaciers and snowfields show that, without exception for the 96 years of record, three or more consecutive years of snowfall above a certain value always resulted in a thickening or advance of those glaciers under study. This was observed six times between 1870 and 1953. Thus, analysis of the long Sierra snowfall record offers a rare opportunity to quantify one climatologic parameter favorable to glacier growth for small alpine cirque glaciers and to gain insight into the general character of what is generally referred to as a "glacial climate."

Since 1870 record was made of the amount of snow that accumulated during each storm at Donner Pass (2200 m), as well as the total depth of the snowpack at various time intervals. Daily snow-depth measurements have been kept since 1897. The actual site of the snow gauge has been changed several times since 1870, but its position has always been within 4 km of the original site in physiographically comparable positions near the broad, gentle pass summit. For the purposes of this analysis the record may be considered as that of a single station. Through analysis of this long record, Heald was able to predict accurately the great snowfalls of 1951–1952 and 1966–1967 (Heald, 1949, p. 65).

Figure 6 is a graphical representation of the Donner Pass seasonal snowfall record. The seasonal snowfall curve has been smoothed twice by the formula $1/4(a + 2b + c)$, where b is the total snowfall for the season in

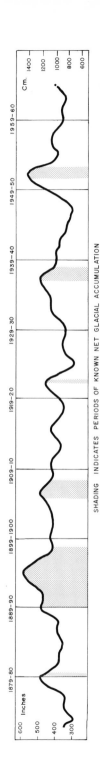

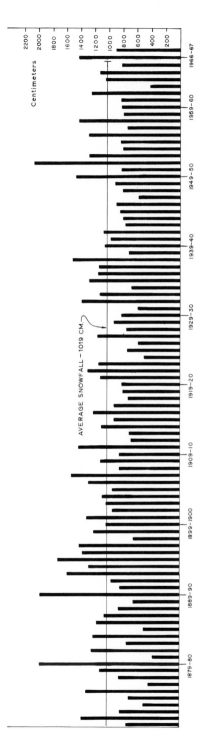

Figure 6. Snowfall at Donner Pass, 1870 to 1966. The curve represents data taken directly from the histogram but smoothed twice according to the formula 1/4(a + 2b + c), where b is the snowfall for the season in question and a and c are the values for the preceding and following seasons.

question, a is the snowfall for the preceding season, and c is the snowfall for the succeeding season. These data were then plotted on probability paper in the manner described by Leopold (1959, p. 4). Such a plot permits one to estimate the frequency of events of greater or lesser magnitude than those observed during the available record. It is understood that this analysis is meaningful only where maritime influence on air masses is such that snowfall depth and water equivalence have a near 1:1 relation, as found in the major western-slope Sierran winter storms.

Figure 7 represents a probability plot for 3-year cumulative snowfall at Donner Pass for the 96 years of continuous record. A perfectly random sequence of events should plot as a straight line on arithmetic-probability paper. As can be seen from the two lines drawn through the points represented on Figure 7, these data are best represented by a smooth curved line. This

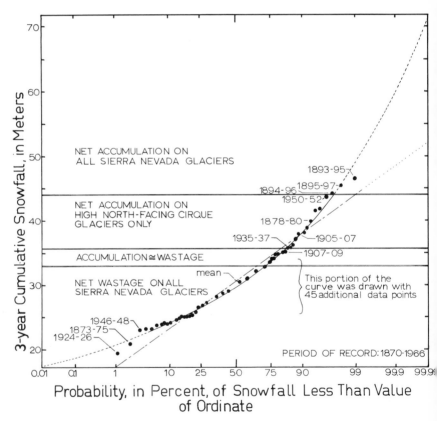

Figure 7. Probability plot of 3-year snowfall at Donner Pass. Where dates are noted, the date refers to the year of the spring of the season in question.

nonrandomness is noted in much hydrologic data and can be explained as a tendency for grouping, termed persistence (Leopold, 1959, p. 8). In the case of the Donner Pass snowfall, the curved arithmetic-probability plot implies that consecutive 3-year periods of above-average snowfall occur more often than they should and are not randomly distributed. Also implicit in the shape of this curve is the fact that fewer periods of very low snowfall have occurred than would be expected if total seasonal snowfall were random. In other words, the frequency of above-average snowfall is greater than one would expect and the frequency of below-average snowfall is less than one would expect. Thus, the occurrence of a markedly heavy snowfall one season indicates that there is an increased chance for above-normal snowfall in the succeeding season, but a dry season does not indicate an increased chance of dryness for the next season. Wet years occur in groups more often than dry years occur in groups.

When the net mass budgets of Sierran glaciers are compared with the 3-year cumulative Donner Pass snowfall, a remarkable uniformity of correlation is found. Systematic mass budget surveys of many Sierran glaciers have been made since 1930 by National Park Service personnel and private individuals (Harrison, 1951, p. 78; Heald, 1947, p. 333; Matthes, 1948, p. 87; unpublished reports of the Park Naturalist, Yosemite National Park, 1934ff). Mass budgets before 1930 may be determined semiquantitatively from analysis of old photographs. Photographs, written accounts, or glacier-survey records showing conditions on one or more Sierran glaciers at comparable times of the summer season were found for at least some portion of the High Sierra for all but 9 years between 1881 and 1966. For 1870 through 1881, some photographs are available, but they are not annotated carefully enough for more than general use for conditions during the decade. Sufficient notes by John Muir and newspaper reports are available for that period, however, to indicate that it was one of marked diminution of glaciers and snowfields until the great snowfall of 1879–1880. During the summer of 1880, snow remained on mountain peaks, blocking passes and mining tunnels, for two months longer than had ever been known to miners and sheep ranchers, few of whom could be expected to have lived there for more than 20 years. Carefully annotated photographs taken by I. C. Russell indicate that, by August and September 1881, snow conditions in the High Sierra were as dry as those of almost any succeeding late summer. References to all source materials and interpretations made therefrom for each year of record are listed elsewhere (Curry, 1968a, App. 1).

When conditions of advance, retreat, or stillstand were determined for the final year of each of the 3-year overlapping snowfall periods, it was possible to make the generalizations noted on Figure 7 regarding correlation between Donner Pass snowfall and conditions of High Sierra snowfields and glaciers.

Within the limits of available photographs and the ability to take semiquantitative data accurately from these photographs, there were only three exceptions to the generalized correlations drawn in Figure 7. These exceptions occurred when, in winters such as that of 1947–1948, unusually persistent winds concentrated a below-normal snowfall in deep drifts on the glaciers and firn fields, resulting in a measurable thickening of them for several succeeding years. Unusually windy winters with normal to subnormal precipitation commonly are times of destructive wind-slab avalanches. Conditions leading to such avalanches in the Sierra are described in detail by Leonard (1951, p. 137, 1949, p. 74). Such snowslides occur throughout the High Sierra and periods of unusual wind-slab activity may be dated by tree-ring analysis of trees growing in or near lee-slope avalanche tracks through subalpine forests. The three periods of occasional minor glacier growth following dry seasons were found to follow within 3 years of seasons of unusually destructive wind-slab avalanches throughout the High Sierra study region.

It can be seen from Figure 7 that the mean 3-year snowfall for the period of record falls below the value needed for even the maintenance of Sierran glaciers. Thus, on the whole, glaciers have retreated during the period of record; however, in the period 1890 to 1897, old photographs clearly document a rapid advance of existing cirque glaciers up to the moraines of the previous neoglacial advance. In one case, it could be demonstrated that a new glacier re-formed in a site where none had existed in 1881. During the summers of 1894, 1895, and 1896, the photographs of J. N. LeConte indicate that snow remained on the ground throughout much of the High Sierra until at least September in areas that would be snow-free by June 15 to July 1 in a more typical year, such as 1964. Once the glaciers of this 1890 to 1897 period of advance reached the pre-existing moraines, which occurred in all documented cases by the year 1900, the ice remained at this maximum position until the summer of 1907 or 1908, when it began to retreat. Although this ice overtopped the terminal moraines of the earlier latest neoglacial (1300 to 1883 A.D.) advances of the Conness, Dana, and Lyell glaciers in Yosemite National Park, it deposited only a veneer of till, generally less than 1 m thick, on the upstream faces of these composite moraines.

Additional minor advances were noted on some north-facing cirque glaciers following the heavy snows of 1880 to 1881, 1905 to 1907, 1936 to 1937, and 1951 to 1953. Thus, a mean seasonal snowfall for three or more consecutive years at Donner Pass of greater than 14.5 m will result in positive mass budgets of all Sierran glaciers, whereas snowfalls of greater than 12 m will result in advances or thickenings of only some Sierran glaciers.

When one plots a probability curve for the snowfall for each season, more information can be gained. The solid curve in Figure 8 represents the seasonal probability curve for the 96 seasons of available data. Since even the smallest

Sierran glaciers may not advance after only a single season of positive mass budget, generalizations about the state of the glaciers as a function of seasonal snowfall could not be made for the bulk of the points represented on this curve. However, those few seasons of greater than 1530 cm of snowfall at Donner Pass were all followed by glacier advances during the summers of the years indicated.

General correlation of the seasonal snowfall data was made with extant records of the levels of pluvial lakes in California and Nevada. Such lakes reflect the presence or absence of drought rather than the presence or absence of periods of excess cold precipitation as do the Sierran glaciers. Lake records used for this comparison include Pyramid and Winnemucca Lakes in Nevada with source waters at Donner Pass (record: 1840 to 1940; Hardman and Venstrom, 1941, p. 71), Mono Lake, in eastern California with headwaters at

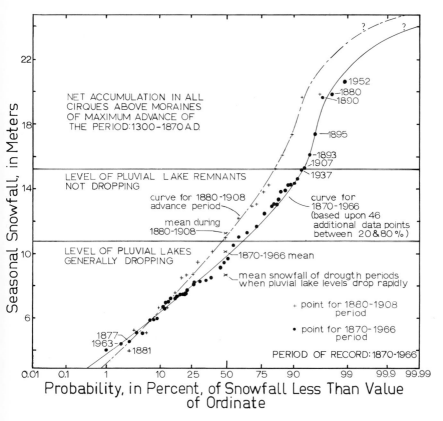

Figure 8. Probability plot of seasonal snowfall at Donner Pass. Where dates are noted, the date refers to the year of the spring of the season in question.

Tioga Pass (record: 1860 to 1940; Thomas, 1962, p. 39), and Lake Elsinore in southern California (record: 1775 to 1930; Lynch, 1931, p. 11). Individual seasons of drought in the Sierra snowfall record were always reflected in lake levels in at least two of the three pluvial lake systems studied, and, although necessarily vague due to the qualitative nature of the lake record, the generalized correlation noted on Figure 8 is correct for all three of the widely separated lake basins for 83 percent of the seasons between 1870 and 1966.

When one compares the mean seasonal snowfalls at Donner Pass during periods of advance and retreat to the mean for the full period of record (1019 cm) and for the 1930 to 1960 seasonal standard climatic normal period (1020 cm), one can quantify the snowfall characteristic of the periods of extremes. The mean for the 1880 to 1908 period of advance from minimum positions of 1879 to the maximum positions of 1895 to 1897 is 1136 cm. This means that an advance can occur during periods when the mean seasonal snowfall is greater than normal by 11 cm, or about one-third the standard deviation for the period of record. This amount of additional snowfall is equivalent to the amount that would fall on Donner Pass during any two days of an average 3-day midwinter or spring storm. Thus, the period 1880 to 1908 differed from that of the full length of record by having, on the average, about one additional storm per season.

The dashed curve on Figure 8 is a probability curve for the period 1880 to 1908 only. Since the slope of the curve during the period of general advance is steeper than that for the whole period of record, we may conclude that there was a greater frequency of cold, wet years and a slightly greater, although not statistically significant, frequency of very dry years. In similar manner, the probability curves for periods of drought, although not plotted in Figure 8, are flatter than that for the total period of record and indicate a narrower range of extremes during general drought periods. The implied theory that the frequency and magnitude of cold, wet winters increase significantly with only a small increase in mean snowfall is supported by the tree-ring record (Fig. 4). During lichenometrically determined periods of glacial advance, the range of tree-ring widths is greater than during periods of known or presumed drought. In like manner, the data for Truckee River from Hardman and Venstrom (1941, p. 71) suggest that the frequency of seasons of high runoff has been greater than average at about the times of Sierran glacial advances during the period 1870 to 1940.

During the period of most rapid glacier advance in 1885 to 1898, mean Donner Pass snowfall was 1323 cm. This is 304 cm more snowfall per season or the amount that would fall, on the average, during two normal 3-day storms. Thus, two additional storms per season for periods of 13 years could account for advances to the limits of the moraines of all periods of advance between 1050 A.D. and the present. To account for the depression of orographic firn limit needed to form the more extensive moraines of 2000 and 2600 years B.P., the same two-storm-per-season snowfall increase would have had to occur for 20 to 60 consecutive winters. This estimate is based upon rough volumetric comparison of the glaciers of each advance period. The 1930 to 1960 average volume of ice in the Conness, Lyell, MacClure, and Dana glaciers of Yosemite National Park would have to be increased by 2 to 2.8 times to equal the volumes of ice estimated to have existed in those same glaciers in photographs taken between 1895 and 1908 at maximum extension to the late neoglacial moraines. If one assumes that this volume increase was due to the increase of two storms per season for 13 years, one may compute an annual rate of firn accumulation that would be associated with the indicated increase in Donner Pass snowfall. The 2000-year-old moraines for the glaciers in Yosemite National Park would require a volume of ice 2 to 4 times as great as that of the late neoglacial maximum, or 4 to 11 times the volume of the glaciers during 1930 to 1960. Thus the average annual accumulation rate of 1885 to 1898 would have to continue from 20 to 60 consecutive years to form glaciers of equal extent to those of 2000 years B.P.

From this analysis, we may conclude that two additional snowstorms per season of average magnitude are sufficient to cause a neoglacial advance equal to those of the 14th through 19th centuries and that one additional storm per season is sufficient to maintain mass budgets of the Sierran glaciers at the positions of their maximum late neoglacial advance. For the total period of record, snowfalls typical of the period of maximum advance occur in only 16 percent of the seasons.

CLIMATIC MODEL FOR THE PLEISTOCENE MAXIMA

Three interrelated questions are raised by the foregoing analysis: (1) How does the climate during a period of advance of small cirque glaciers compare with that of the periods of major glacial advance in the Sierra Nevada? (2) Could one account for the full extent of the ice during the maximum Wisconsin advance by raising the mean seasonal snowfall a mere 300 cm and maintaining that snowfall for sufficient time to accumulate the estimated maximum ice volume? (3) How can one estimate the Donner Pass snowfall typical of the times of maximum Pleistocene glacial advance?

If the snowfall probability curves were linear or could be accurately interpolated beyond the limits of the data for the heaviest snowfall seasons, one could, in theory, predict the extremes of snowfall that would be expected at 1000- or 10,000-year intervals (abscissa values of 99.9 and 99.99 percent, respectively). Unfortunately, the curve in Figure 8 lacks critical data that would establish its true shape during times of very high snowfall. It is unreasonable to assume that a probability curve for snowfall can become asymptotic. If the existing Donner Pass data were not interpolated as shown in Figure 8 but instead were made to fit a single smooth, asymptotic curve, one would be forced to conclude that infinite snowfall occurred at 1000-year intervals. At some point, therefore, the curve must recurve, as is suggested by the three existing data points above 20 m seasonal snowfall. If the upper limits of the seasonal-snowfall probability curve are as shown in Figure 8, one could conclude that the extreme maximum snowfall at 10,000- to 30,000-year intervals was never much more than 24 m, or only 116 percent of the 1952 snowfall. Since the shape of the curve is in doubt, the extremes of snowfall during time intervals equivalent to those of the major Wisconsin advances are also uncertain. However, it is possible to place limits on the value of mean snowfall for the maximum Wisconsin advances by computing the snowfall necessary to lower the firn limit to its elevation during those times.

Wahrhaftig's map of lowest Wisconsin firn limits for the Sierra Nevada (Wahrhaftig and Birman, 1965, p. 304) shows that the climatic firn limit at those times nearly coincided with the flat summit of Donner Pass. That estimate is supported by my observation that the distribution of glaciated bedrock and cirque floors in the region suggests that only a thin, marginal ice cap existed at the pass summit during the Wisconsin maxima. Thus, the maximum Wisconsin snowfall at Donner Pass (2200 m) was just sufficient to maintain a slight net accumulation under the general temperature and precipitation regimen of the time.

By estimating the rate of loss of the spring snowpack at Donner Pass at various periods during the melting season, one can estimate the amount of snow that would have to fall there during the accumulation season so that there would be a slight net annual accumulation. Daily snow depths at the Donner Pass gauging station have been recorded since 1897 and monthly depths at various other snow-survey sites near the Weather Bureau gauging station since 1930. From these data, chiefly reported in the summary volume of the California state snow survey (California Department of Water Resources, 1965), one can compute the average rate of loss of the snowpack by melting, sublimation, and metamorphism for as long as some snowpack exists at the exposed meadow gauging stations. Even in the years of heaviest snowfall, snow has not remained on the ground in these unshaded sites beyond the end of July.

To estimate how fast snow would melt at Donner Pass in August and September if snow were present requires interpolation based upon the shape of the rate of melting curves for higher elevations where snow remains for longer periods of time. Figure 9 illustrates snow-melt curves for the average length of the melting season for various elevations in Yosemite National Park and at Donner Pass. These curves were constructed from snow-survey data and photographs for the period 1930 to 1960 only. Since the snow surveys are not conducted after June, the aerial and ground photographs were used almost exclusively to determine the rate of net decrease of snow depth at areas such as Tuolumne Meadows and Tioga Pass. The resulting curves of Figure 9 are only approximate due to the nature of the data. The range of variation of rate of melting between cloud-free, hot summers and overcast, cool summers was more than 100 cm per month. Thus, at Donner Pass, a given depth of snow would last twice as long during a cool summer as during a year with a hot, cloud-free spring.

From Figure 9 we may estimate that the spring snowpack on April 15 at 2400 m would have to be 850 to 900 cm deep for snow to persist to the beginning of the next accumulation season under the average summer climatic conditions of 1930 to 1960. Analysis of the full 1897 to 1966 snow-depth record for Donner Pass shows that the April 1 snowpack there is equal to about one-third of the total seasonal snowfall. Thus, at 2400 m near Donner Pass, the mean snowfall during the Wisconsin maxima would have to be 25 to 27 m, or 2.5 times the present mean snowfall, assuming that summer cloud cover, ground albedo, and temperatures during the glacial maxima were the same as those of 1930 to 1960. With cool, cloudy summers such as those of 1895 and 1907, a mere 12 m of winter snowfall could be expected to last

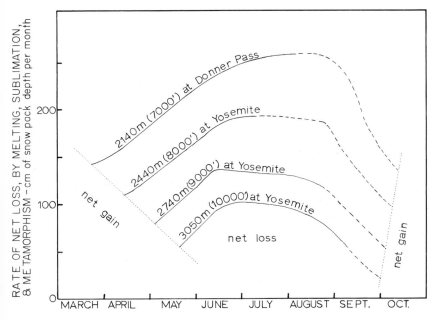

Figure 9. Average rates of net reduction of the spring and summer snowpack for the High Sierra for the period 1931 to 1960.

through the summer at 2400 m if one assumes that the rate of melting at such times was 100 cm per month less than normal. Since the photographic record shows that cool, cloudy summers generally follow winters of above-average snowfall, possibly due in part to a positive feedback mechanism related to the increased albedo of the snow-covered mountains, it is reasonable to assume that mean seasonal Donner Pass snowfall during glacial maxima was between 12 and 27 m.

Photographs show that in the summers of both 1895 and 1907 some snow remained on the ground at Tioga Pass at 3000 m elevation throughout the summer. These summers were preceded by Donner Pass snowfalls of 17 and 15 m, respectively. In the warm, cloud-free summer of 1952, after a snowfall of more than 20 m at Donner Pass, snow was gone from the ground at Tioga Pass by the end of August. Since the maximum observed mean seasonal snowfall at Donner Pass during the 1885 to 1898 advance was only 13 m, the above estimates of snowfall based upon rate of snow melt suggest that climatic conditions during periods of neoglacial advance were not as severe as those during maximum Wisconsin glacial stages. The 13-m mean snowfall would be just barely sufficient to lower the climatic firn limit to its lowest Wisconsin limit if one assumes that the midglacial summers were as cool and cloudy as those of the extremes of the historical record. Such an assumption is possible, especially when considering that the average albedo of the entire High Sierra for June through September of 1907 was probably about 80 percent as compared to 30 percent during an average modern summer such as that of 1964. (Albedo estimates were made by noting snow cover in old photographs taken from mountain summits in the Sierra and comparing them with data published by McFadden and Ragotzkie [1967, p. 1135].)

The conclusions of this analysis of snowfall data are that mean seasonal snowfall, which is a function of both precipitation and winter temperature, would have to be increased by 20 to 30 percent for periods of 13 to 60 years to account for all glacial advances known to have occurred during the last 2600 years. It is doubtful, although possible, that mean snowfalls only 30 percent above normal also characterized the times of the Wisconsin maxima. Most probably these major glacial maxima were associated with snowfalls less than 2.5 times the 1930 to 1960 mean, and, if one visualizes a positive feedback mechanism where summers become cooler and cloudier as glaciers expand in extent at a constant mean snowfall, it is possible that snowfall was no more than 1.5 times that of the present climatic normal during the Wisconsin maxima. The chief finding of this analysis is that the law of uniformitarianism could apply to the known Pleistocene climatic fluctuations. Mean climatic conditions known to have occurred for short periods within the 96 years of record could explain the maximum depression of the Wisconsin firn limit if such conditions occurred for longer periods of time.

GENERAL PALEOCLIMATIC SEQUENCE

The results of this study are summarized in a chart showing climatic variation as a function of time. Since snowfall variation is thought to reflect the differences between glacial and interglacial climates in the High Sierra, the mean seasonal snowfall has been plotted along the abscissa of the paleoclimatic curve (Fig. 10). The actual shape of the curve during the periods of glacial advance has been estimated by calculating both the mean snowfall and the number of years it must have prevailed to create glaciers of the sizes mapped for the various advances in several drainage basins while assuming summer climates typical of those of the period 1895 to 1908. These calculations, while involving many uncertainties, yield the general conclusion that mean snowfalls less than 50 percent above normal could account for the volumes of glacial ice that are presumed to have formed during the various advances of the last 10,000 years. Firn limit and ice area and volume calculations are based upon my own mapping of the glacial deposits of the Mammoth, Convict, and Deadman Creek drainages on the eastern side of the Sierra and on the glacial map of the drainages of Rock Creek and Mono Creek on both sides of the range crest by Birman (1964, Pl. 1). Correlation of Birman's glacial deposits with those in other valleys and with the charted climatic events is done on the basis of lichenometry, dendrochronometry, and general glacial stratigraphy. The basis of these correlations will be the subject of a chapter in a later paper on the glacial history of the Mammoth Lakes region of the Sierra Nevada.

CORRELATION BEYOND THE SIERRA NEVADA

The generalized climatic curve (Fig. 10), based entirely on data derived wholly from the state of California and Truckee River drainage in Nevada, is similar to some climatic curves constructed for other parts of the world.

Most other authors base their glacial and inferred paleoclimatic curves directly and indirectly on radiocarbon dates. The latest work on the accuracy of radiocarbon supports the conclusion that C^{14} dating of uncontaminated terrestrial wood is accurate to within about 200 years back to approximately 2000 years B.P. (Damon and others, 1966, p. 1057). Between 2000 and at least 7000 years B.P., radiocarbon dates are commonly 500 to 1000 years too young. Thus the climatic curve for the Sierra Nevada can be compared directly with those of other areas for the period 2600 to 10,000 years B.P. and may be considered as roughly comparable to radiocarbon-based chronologies for the period 0 to 2600 years B.P. within an expectable variation of ± 200 years.

A rather complete summary of the known data on glacial fluctuations for the mountainous regions of all of North America has just been completed

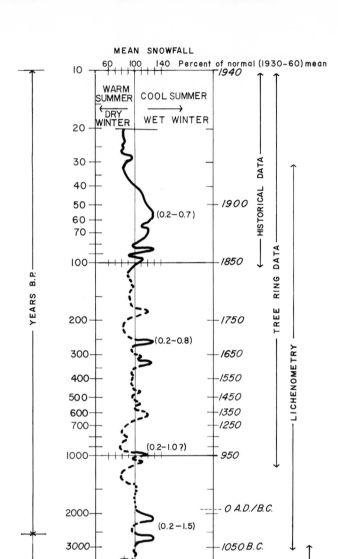

Figure 10. Paleoclimatic curve representing variations in Sierra Nevada snowfall during the last 10,000 years. Time is plotted on a logarithmic scale. Figures in parentheses indicate the range, in percent, of the relative magnitudes of the glacial advances occurring at the various times as a function of the maximum volumes of glaciers that existed during the time of the late Wisconsin maximum.

by Porter and Denton (1967, p. 177). These authors conclude that fluctuations within at least the last 3000 years appear contemporaneous throughout North America and the rest of the world. Their data from world-wide sources lends support to the existence of Sierran advances of 2000 to 2600 years B.P. and 1300 to 1900 A.D. (Porter and Denton, 1967, p. 201). The period of Sierran advance that occurred about 1000 years B.P. is well-represented in Porter and Denton's data by a cluster of radiometric dates directly associated with glacial advances or surges in Alaska and with one date on an aggraded river terrace in the southwestern United States (1967, p. 199). The least disputable evidence cited by Porter and Denton for Australia, New Zealand, and South America also supports the time of the most recent period of Sierran advance as well as that of 2000 to 2600 years B.P.

J. B. Benedict in the Colorado Rocky Mountains has recently radiometrically dated organic silt taken directly from glacial ice making up the core of a moraine that bore a lichen cover that indicated that it was formed between the last advance of the period 2000 to 2600 and 900 years B.P. This silt yielded an age of 1000 ± 90 b.p. (I-2562) and thus appears to support the existence of the brief period of glacial climate noted in the Sierra at about that time (Benedict, 1966, personal commun.).

In the Austrian Alps Heuberger (1966) and Mayr (1964) have recently established an unusually complete chronology of the glacial history of the last 10,000 years. Their data, based upon both radiocarbon and lichenometric dates, yielded limiting data on glacial maxima that occurred between approximately 7500 and 6000 years B.P., 2900 and 2000 years B.P., and 1600 and 1920 A.D. Thus, these Alpine data support all the Sierran advances except that of about 1000 years B.P. Recent Soviet work claims to have recognized advances contemporaneous with all of the recognized Sierran advances of the last 10,000 years and three additional periods of advance at 4000, 5200 to 6000, and 8500 to 9600? b.p. (Maximov, 1966, p. 84). Both the Soviet and Austrian Holocene glacial advances were of approximately the same size as the correlative Sierran advances relative to the maximum of the latest Wisconsin or Würm advance.

In summary, available data suggest that world-wide, temperate-latitude glacier advances occurred nearly synchronously during the last 10,000 years. Synopses of known historical data on climatic fluctuation, such as those published by Lamb and Johnson (1961, p. 398) and Betin and Preobrazenskij (1962), present general climatologic data for North Africa, Europe, and the western Soviet Union that show that periods of climatic extremes generally occur in those areas within 20 years of a similar occurrence in the Sierra Nevada. For instance, the extraordinarily severe winters of 1608, 1621, and 1624 to 1634 in Europe parallel periods of cool, wet Sierran climate as reflected in tree rings. However, the magnitudes of the climatic changes, and the resulting glacier advances or retreats, vary widely during a given fluctuation throughout the world. The Sierran glacial advances of about 1000 years B.P. reflect only a brief period of heavy snowfall with advances in some Sierran cirques, whereas in Alaska this was a period of sizable glacial advance or surge. At the same time multiple glaciations appear to have occurred in the Colorado Rockies and the Soviet Union, severe winters occurred in the Mediterranean countries with freezing of the Nile between 764 and 860 and again in 1011 A.D., and yet no glacial deposits are known to have been reported for this period from the Alps or Scandinavia.

CONCLUSIONS

Diverse techniques of paleoclimatic analysis, such as lichenometric dating of high-altitude mass-wasting features, limited analyses of tree-ring width variations, study of historical photographs, and analyses of age classes of vegetation, can be integrated into one paleoclimatic curve with much better agreement among techniques than had been expected when this research was first undertaken. If one adopts a model of the climate of times of general glacial advance as one in which a 40- to 50-percent increase in seasonal snowfall is associated with cooler, cloudier summers, it is possible to say that all of the climatic variations thought to have occurred in the Sierra Nevada during the last 10,000 years could have been the result of changes in mean climatic parameters that are within the range of known variation in the extremes of those parameters during the last 100 years.

Geologic data as well as analyses of various climatic models suggest that at least three of the four or five periods of post-Wisconsin glacial advance noted in the Sierra Nevada were characterized by intermittent advances and retreats. Statistical analyses of 96 years of snowfall data encompassing the period of a single small neoglacial advance suggest that the driest, although not necessarily warmest, recorded individual seasons may have occurred during the general periods of glacial advance. These snowfall data suggest that short periods of high climatic variability are associated with periods of above-average mean precipitation while periods of low year-to-year climatic variability are associated with periods of average or below-average precipitation.

Deglacial climates are typified by the general circulation pattern of the 1930 to 1960 climatic normal period, whereas the so-called "climatic optimum" was in reality two or more periods of generally decreased wavelengths of westerly circulation of the upper atmosphere when storm tracks remained in narrow belts at high latitudes. Such conditions may have been accompanied by an increase in summer thundershower activity in the Sierra Nevada as moist tropical air masses migrated farther north than they did during periods of glacial advances. Glacial climates apparently characterized times of increased vigor of upper-atmosphere circulation with higher amplitude, longer wavelength upper-atmosphere meander patterns and resultant more frequent frontal storms and more southerly extension of the storm tracks.

The use of tree rings, lichenometry, and vegetational-age-class analyses for the interpretation of past climatic variations was found to be possible only when the complex interactions between the organisms studied and their presumed environments were appreciated.

REFERENCES CITED

Adam, D. P., 1967, Late-Pleistocene and Recent palynology in the central Sierra Nevada, California, p. 275–301 in Cushing, E. J., and Wright, H. E., Jr., Editors, Quaternary paleoecology: New Haven, Connecticut, Yale Univ. Press, 433 p.

Adamenko, V. N., 1963, On the similarity in the growth of trees in northern Scandinavia and in the polar Ural Mountains: Jour. Glaciology, v. 4, p. 449–451.

Andrews, J. T., and Webber, P. J., 1964, A lichenometrical study of the northwestern margin of the Barnes Ice Cap: a geomorphological technique: Ottawa Geog. Bull., no. 22, p. 80–104.

Antevs, Ernst, 1925, On the Pleistocene history of the Great Basin: Carnegie Inst. Washington Pub. 352, p. 51–144.

Bateman, P. C., and Wahrhaftig, Clyde, 1966, Geology of the Sierra Nevada, p. 107–172 in Bailey, E. H., Editor, Geology of northern California: California Div. Mines and Geology Bull. 190, 508 p.

Benedict, J. B., 1967, Recent glacial history of an alpine area in the Colorado Front Range, U.S.A., Part I. Establishing a lichen growth curve: Jour. Glaciology, v. 6, p. 817–832.

—— 1968, Recent glacial history of an alpine area in the Colorado Front Range, U.S.A., Part II. Dating the glacial deposits: Jour. Glaciology, v. 7, p. 77–87.

Beschel, R. E., 1961, Dating rock surfaces by lichen growth and its application to glaciology and physiology (lichenometry), p. 1044–1062 in Raasch, G. O., Editor, Geology of the Arctic, v. II (Proc. First Internat. Symposium on Arctic Geology): Toronto, Toronto Univ. Press.

—— 1965, Epipetric succession and lichen growth rates in the eastern neararctic, p. 25–26 in Abstracts, 7th INQUA Congress: Boulder, Colorado, 532 p.

Betin, V. V., and Preobrazenskij, Ju. V., 1962, The severity of winters in Europe and the state of ice cover in the Baltic: Leningrad, Gidrometeoizdat, p. 18–71.

Birman, J. H., 1964, Glacial geology across the crest of the Sierra Nevada, California: Geol. Soc. America Spec. Paper 75, 80 p.

Bray, J. R., 1965, Forest growth and glacier chronology in northwest North America in relation to solar activity: Nature, v. 205, p. 440–443.

—— 1966, Similarity of tree growth in northern Scandinavia, polar Urals and the Canadian Rockies: Jour. Glaciology, v. 6, p. 321–322.

Bray, J. R., and Struik, G. J., 1963, Forest growth and glacier chronology in eastern British Columbia, and their relation to recent climatic trends: Canadian Jour. Botany, v. 41, p. 1245–1271.

California Department of Water Resources, 1965, Snow survey measurements through 1964: California Dept. Water Resources Bull. 129, 366 p.

Chamberlin, T. C., 1897, A group of hypotheses bearing on climatic changes: Jour. Geology, v. 5, p. 653–683.

Clausen, Jens, 1965, Microclimatic and vegetational contrasts within a subalpine valley: Natl. Acad. Sci. Proc., v. 53, p. 1315–1319.

Curry, R. R., 1966, Glaciation about 3,000,000 years ago in the Sierra Nevada: Science, v. 154, p. 770–771.

—— 1968a, Quaternary climatic and glacial history of the Sierra Nevada, California: Ann Arbor, Michigan, University Microfilms, Order No. 68-13,896, 238 p.

—— 1968b, California's Deadman Pass glacial till is also nearly 3,000,000 years old: California Div. Mines and Geology Mineral Inf. Service, v. 21, no. 10, p. 143–145.

Dale, R. F., 1959, Climates of the states—California: U.S. Weather Bureau, Climatography of the U.S., no. 60-4, 46 p.

Dalrymple, G. B., 1969, Potassium-argon ages of Recent rhyolites of the Mono Craters, California: Earth and Planetary Sci. Letters, in press.

Damon, P. E., Long, Austin, and Grey, D. C., 1966, Fluctuation of atmospheric C^{14} during the last six millennia: Jour. Geophys. Research, v. 71, p. 1055–1063.

Daubenmire, R. F., 1954, Alpine timberlines in the Americas and their interpretation: Butler Univ. Bot. Stud., v. 11, p. 119–136.

Douglass, A. E., 1936, Climatic cycles and tree growth: Carnegie Inst. Washington Pub. 289, v. 3, 171 p.

Ferguson, C. W., Huber, B., and Suess, H. E., 1966, Determination of the age of Swiss Lake dwellings as an example of dendrochronology-calibrated radiocarbon dating: Zeitschr. Naturforschung, Ausgabe A: Astrophysik, Physik und Chemie, v. 21a, p. 1173–1177.

Flint, R. F., 1957, Glacial and Pleistocene geology: New York, John Wiley & Sons, Inc., 553 p.

Fritts, H. C., 1965a, Tree-ring evidences for climatic changes in western North America: Monthly Weather Rev., v. 93, p. 421–443.

—— 1965b, Dendrochronology, p. 871–879 in Wright, H. E., Jr., and Frey, D. G., Editors, The Quaternary of the United States: Princeton, New Jersey, Princeton Univ. Press, 922 p.

—— 1966, Growth-rings of trees: their correlation with climate: Science, v. 154, p. 973–979.

Hale, M. E., Jr., 1961, Lichen Handbook: Smithsonian Inst. Pub. 4434, 178 p.

Hardman, George, and Venstrom, Cruz, 1941, A 100-year record of Truckee River runoff estimated from changes in levels and volumes of Pyramid and Winnemucca Lakes: Trans. Am. Geophys. Union, pt. 1, p. 71–90.

Harrison, A. E., 1951, Are our glaciers advancing?: Sierra Club Bull., v. 36, no. 5, p. 78–81.

Heald, W. F., 1947, Palisade Glacier survey, Sierra Nevada: Am. Alpine Jour., v. 6, no. 3, p. 333–336.

—— 1949, Sierra snows—past and future: Sierra Club Bull., v. 34, no. 6, p. 55–67.

Heuberger, Helmut, 1966, Gletschergeschichtliche Untersuchungen in den Zentralalpen zwischen Sellrain- und Ötztal: Innsbruck, Universitats-verlag Wagner, Wissenschaftliche Alpenvereinshefte, v. 20, 126 p.

Huntington, Ellsworth, 1925, Tree growth and climatic interpretations: Carnegie Inst. Washington Pub. 352, p. 145–212.

Kauffman, E. G., and McCulloch, D. S., 1965, Biota of a late glacial Rocky Mountain pond: Geol. Soc. America Bull., v. 76, p. 1203–1232.

LaMarche, V. C., and Mooney, H. A., 1967, Altithermal timberline advance in western United States: Nature, v. 213, p. 980–982.

Lamb, H. H., and Johnson, A. I., 1961, Climatic variation and observed changes in general circulation, Pt. III: Geog. Annaler, v. 43, p. 363–400.

Lamb, H. H., Lewis, R. P. W., and Woodroffe, A., 1966, Atmospheric circulation and the main climatic variables between 8000 and 0 B.C.: meteorological evidence, p. 174–217 in Sawyer, J. S., Editor, World climate from 8000 to 0 B.C.: London, Royal Meteorological Society, 229 p.

Lange, O. L., 1965, Der CO_2-Gaswechsel von Flecten bei tiefen Temperaturen: Planta, v. 64, p. 1–19.

Leonard, R. M., 1949, Echo Lake avalanches: Sierra Club Bull., v. 34, no. 6, p. 74–81.

—— 1951, Echo Lake avalanches, 1950: Sierra Club Bull., v. 36, no. 5, p. 137–140.

Leopold, L. B., 1959, Probability analysis applied to a water-supply problem: U.S. Geological Survey Circ. 410, 18 p.

Lynch, H. B., 1931, Rainfall and stream runoff in southern California since 1769: Los Angeles, Metropolitan Water Dist. of Southern Calif., 31 p.

Løken, O. H., and Andrews, J. T., 1966, Glaciology and chronology of fluctuations of the ice margin at the south end of the Barnes Ice Cap, Baffin Island, N.W.T.: Geog. Bull. (Ottawa), v. 8, no. 4, p. 341–359.

Mann (the Abbé Mann), 1779, On the gradual changes in temperature and soil which take place in different climates, with an enquiry into the cause of those changes: Philos. Mag., v. 4, p. 337–347; v. 5, p. 18–23.

Matthes, F. E., 1940, Report of the committee on glaciers, 1939–40, pt. 2, section on hydrology: Am. Geophys. Union Repts. and Papers, p. 396–406.

—— 1948, Moraines with ice cores in the Sierra Nevada: Sierra Club Bull., v. 33, no. 3, p. 87–96.

Maximov, E. V., 1966, Absolutina khronologia stadii sokrashenia gornikh lednikov: Sov. Geologia, no. 3, p. 84–96.

Mayr, Franz, 1964, Untersuchungen über Ausmass und Folgen der Klima- und Gletscherschwankungen seit dem Beginn der Postglazialen Wärmezeit: Zeitschr. Geomorphologie, v. NF 8, p. 257–285.

—— 1968, in Richmond, G. M., Editor, Quaternary glaciation in the Alps: Boulder, Colorado Univ. Press.

McCulloch, David, and Hopkins, David, 1966, Evidence for an early Recent warm interval in northwestern Alaska: Geol. Soc. America Bull., v. 77, p. 1089–1108.

McFadden, J. D., and Ragotzkie, R. A., 1967, Climatological significance of albedo in Central Canada: Jour. Geophys. Research, v. 72, p. 1135–1143.

Miller, D. H., 1955, Snow cover and climate in the Sierra Nevada, California: California Univ. Pubs. Geography, v. 11, p. 1–218.

Munz, P. A., and Keck, D. D., 1963, A California flora: Berkeley, California Univ. Press, 1681 p.

Neustadt, M. I., 1967, The lower Holocene boundary, p. 415–425 *in* Cushing, E. J., and Wright, H. E., Jr., *Editors,* Quaternary paleoecology: New Haven, Connecticut, Yale Univ. Press, 433 p.

Nilsson, Tage, 1965, The Pleistocene-Holocene boundary and the subdivision of the late quaternary in southern Sweden: Rept. of the VIth Intern. Cong. on Quaternary, Warsaw, 1961, v. 1: Subcommission on the Holocene, p. 479–494.

Oosting, H. J., 1958, The study of plant communities, 2nd edition: San Francisco, W. H. Freeman, 430 p.

Pace, Nello, 1963, Climatological data summary for the decade 1 January 1953 through 31 December 1962 from the Crooked Creek Laboratory (10,150 feet) and the Barcroft Laboratory (12,470 feet): Berkeley, California Univ. White Mountain Research Station, 54 p.

Porter, S. C., and Denton, G. H., 1967, Chronology of neoglaciation in the North American Cordillera: Am. Jour. Sci., v. 265, p. 177–210.

Richmond, G. H., 1965, Glaciation of the Rocky Mountains, p. 217–230 *in* Wright, H. E., Jr., and Frey, D. G., *Editors,* The Quaternary of the United States: Princeton, New Jersey, Princeton Univ. Press, 922 p.

Sawyer, J. S., *Editor,* 1966, World climate from 8000 to 0 B.C.: London, Royal Meteorological Society, 229 p.

Schulman, Edmund, 1956, Dendroclimatic changes in semiarid America: Tucson, Arizona Univ. Press, 142 p.

Schove, D. J., 1954, Summer temperatures and tree-rings in northern Scandinavia A.D. 1461–1950: Geog. Annaler, v. 36, p. 40–80.

Smith, G. I., 1968, Late Quaternary geologic and climatic history of Searles Lake, southeastern California: Salt Lake City, Utah Univ. Press.

Stone, E. C., 1962, Climate and vegetation of the upper Sierra, p. 14–22 *in* Proc. May, 1962, conference: Research and Land Management in the Upper Sierra, Berkeley, Wildlands Research Center of Univ. California, 148 p.

Tadeucci, Adiano, and Broecker, Wallace, 1969, Thorium-230 dating of Recent volcanic rocks: Earth and Planetary Sci. Letters, in press.

Thomas, H. E., 1962, The meteorologic phenomenon of drought in the Southwest: U.S. Geol. Survey Prof. Paper 372-A, 42 p.

Thomas, H. E., and others, 1963, Effects of drought in basins of interior drainage: U.S. Geol. Survey Prof. Paper 372-E, 50 p.

U.S. Army Corps of Engineers, 1957, Ten-year storm precipitation in California and Oregon coastal basins: Sacramento, California, Civil Works Investigations Project CW-151, Flood Volume Studies—West Coast, U.S. Army Corps Engineers, Tech. Bull. no. 4, Sacramento Dist., 7 p.

U.S. Weather Bureau, 1932, Climatic summary of the United States, Section 18— Southern California and Owens Valley: Washington, D.C., 41 p.

Wahrhaftig, Clyde, and Birman, J. H., 1965, The Quaternary of the Pacific Mountain System in California, p. 299–340 *in* Wright, H. E., Jr., and Frey, D. G., *Editors*, The Quaternary of the United States: Princeton, New Jersey, Princeton Univ. Press, 922 p.

Wardle, P., 1965, A comparison of alpine timberlines in New Zealand and North America: New Zealand Jour. Botany, v. 3, p. 113–135.

Williamson, Hugh, 1770, An attempt to account for the change of climate which has been observed in the Middle Colonies in North America: Am. Philos. Soc. Trans., v. 1, p. 336–345.

PRESENT ADDRESS: DEPARTMENT OF GEOLOGY, UNIVERSITY OF MONTANA, MISSOULA

GEOLOGICAL SOCIETY OF AMERICA, INC.
SPECIAL PAPER 123

Glacial Ice-Contact Rings and Ridges

RICHARD R. PARIZEK

Pennsylvania State University, University Park, Pennsylvania

ABSTRACT

Ice-contact rings and ridges are minor landforms associated with end moraines, hummocky and transitional ground moraines, hummocky and pitted outwash plains, moraine-lake plateaus, moraine plateaus, former ice-walled and ice-floored valleys, and other glacial landforms. They were investigated in south-central Saskatchewan and are known to occur in the Canadian provinces of Saskatchewan, Alberta, and Manitoba and in the United States in northern Montana, North Dakota, Minnesota, Wisconsin, Illinois, and Indiana.

They are conspicuous landforms on aerial photographs. They resemble rings that are several hundred feet to several miles in diameter and a few tens of feet in relief. The circular features appear as oval, crenulated, or irregular-shaped landforms that form rims on the margins of moraine-lake plateaus and moraine plateaus and irregular regions of hummocky ground moraine. Ice-contact ridges may appear as straight ridges a few hundred feet to several miles long or as irregular, sinuous, or anastomosing ridges and elongate, straight or sinuous ridges that resemble links in a chain. Several ridges may be parallel in certain parts and merge or branch into others. Most are sharp-crested and about 100 to 250 feet across at their base.

Ice-contact rings and ridges are interpreted as the end product of wasting stagnant- or dead-ice masses of various sizes and may be used to determine the manner and history of ice retreat and the most probable origin of associated landforms and deposits. The rings and ridges may form in either glaciolacustrine, glaciofluvial, ablational, or superglacial environments, some by sloughing off of ablational and englacial debris from stagnant-ice blocks and others possibly in part or entirely by a subglacial ice-press mechanism, or a combination of the two. The nature, proposed mechanism, and environment of formation of diverse ice-contact rings and ridges are illustrated. These

49

include ice-contact rings marginal to moraine-lake plateaus and moraine plateaus; ice-contact ridges marginal to partly buried subglacial melt-water channels where ice, till, and/or bedrock served as valley walls; ridges marginal to partly buried melt-water channels where stagnant-ice blocks persisted in melt-water channels after adjacent uplands were deglaciated; ridges marginal to open melt-water channels and outwash plains; ice-contact rings and ridges associated with valleys eroded in ice and largely floored by ice; and ice-contact ridges marginal to hummocky and transitional ground moraine and some lake basins. They have been mapped with washboard moraines, ridged end moraines, and contorted bedrock ridges. Their occurrences with these landforms do not require unique explanations, however.

CONTENTS

INTRODUCTION

I first proposed the terms "ice-contact rings" and "ice-contact ridges" in my unpublished doctoral dissertation in 1961. They were applied to a variety of elongate and circular ridges of low relief classified and mapped in the Willow Bunch Lake area of Saskatchewan (Fig. 1).

I have observed similar features elsewhere in the Canadian provinces of Saskatchewan, Alberta, and Manitoba and in the United States in northern Montana, North Dakota, Minnesota, Wisconsin, and Illinois.

Similar landforms had been reported earlier by Gravenor and Kupsch (1959) in Saskatchewan and Alberta, Stalker (1960) in Alberta, and Hoppe (1952, 1957, 1959, 1963) in the interior of Norbotten, Sweden.

Identical landforms referred to by different names have been mapped by investigators working in North Dakota. Investigators include Clayton, who worked in Logan and McIntosh Counties (1962) and on the Missouri Coteau in North Dakota (1967), Winters (1963) in Stutsman County, Bluemle (1965) in Eddy and Forster Counties, Kume and Hansen (1965) in Burleight County, Hansen (1967) in Divide County, Royse and Callender (1967) in Mountrail County, Pettyjohn (1967) in Ward County, and Bluemle and others (1967) in Wells County. Clayton and Cherry (1967) reported similar landforms in North Dakota and Saskatchewan and Harrison (1963, p. 37–40) in Indiana.

These features undoubtedly have widespread distribution in other areas as well where masses of glacial ice stagnated during recession.

The terms "ice-contact rings" and "ice-contact ridges" are used in a descriptive sense as a convenience in mapping and are applied to a group of landforms that are interpreted to have had an ice-contact origin. They may be associated with end moraines, hummocky and transitional ground moraines, outwash plains, melt-water channels, glacial-lake basins, moraine-lake plateaus, moraine plateaus, ridged end moraines, washboard moraines, and con-

Figure 1. Location of the Willow Bunch Lake area.

torted bedrock ridges (Parizek, 1964, Pl. 3). Their associations with selected glacial landforms and deposits in the Willow Bunch Lake area are shown on Plate 1. Because ice-contact rings and ridges occur with a variety of landforms and deposits it is only reasonable to presume that the mechanics of their formation, even if similar in gross aspect, must have varied in detail as their specific environment of formation varied.

Ice-contact rings and ridges are minor topographic landforms when compared to many other glacial features. They are very useful in a genetic sense, however, because they are clearly derived from wasting stagnant- or dead-ice blocks of various sizes. For this reason, they are diagnostic criteria that can be used to determine the manner and history of ice retreat and aid in the interpretation of the origin of associated landforms and deposits.

General Expression and Composition

Ice-contact rings and ridges are conspicuous on aerial photographs because of their ridgelike or eskerin aspect, by the fact that they may be from 100 feet to several miles in length, and by their low relief and tonal contrast with adjacent areas.

The rings are several hundred feet to several miles in diameter and are oval or crenulated or irregularly shaped. They may encircle moraine plateaus and moraine-lake plateaus, partly border isolated areas of hummocky ground moraine, or appear with hummocky end moraines. Ice-contact rings several hundred feet in diameter and a few tens of feet in relief may display braided, irregular, sinuous, or meandering patterns that extend from several hundred feet to several miles in length. In this occurrence ice-contact rings appear beadlike in that they resemble links in a chain or chain coral. These in turn may merge with more linear ice-contact ridges or give rise to individual rings, dimpled knobs or hummocks, crenulated knobs, or knobs that resemble barley grains.

Ice-contact ridges may be straight, irregular, sinuous, anastomosing, or beadlike where they merge into chainlike ice-contact rings. Several ridges may run parallel and then merge into a single ridge or give rise to a number of other parallel or branching ridges. They vary from a few feet to nearly 200 feet in relief, from several hundred feet to more than five miles in length, and from 200 to 300 feet across at their base. Ice-contact ridges are normally sharp-crested and commonly lag boulders and cobbles cover their surfaces. Their crests are uneven in elevation, and frequently ridges are interrupted by kettle holes or gaps or obscured by knob-and-kettle topography. On photographs their regional continuity normally is apparent, however.

Ice-contact ridges may terminate at the base of hanging melt-water channels or merge into melt-water channels. They may appear as ridges along the borders of open melt-water channels or partly buried melt-water channels, along the margins of hummocky and pitted outwash plains, and along feeder channels leading to outwash plains. They may partly border hummocky ground moraine (dead-ice moraine) or hummocky lake plains where super-glacial lake sediments have collapsed to produce characteristic knob-and-kettle or hummocky topography.

Composition of ice-contact rings and ridges differs, depending upon whether they were formed in glaciolacustrine, glaciofluvial, ablational, or subglacial environments or in combinations of two or more of these environments. Usually, ice-contact ridges are composed of beds of dirty sand and gravel and boulders that commonly lack apparent bedding or that display contorted bedding or angle-of-repose bedding. Till containing a greater abundance of cobbles and boulders when compared to adjacent lodgement-type till is common in some exposures, and in others till resembling that contained in adjacent ground moraine is common. Contorted beds of laminated silt and clay characterize other exposures.

All combinations of till, sand, gravel, boulders, and laminated silt and clay have been observed in other exposures. Individual beds in these associations frequently are highly contorted, even overturned and offset by high-angle faults.

Ice-contact rings are composed principally of till and/or "washed" till more nearly resembling glaciofluvial sediments. Beds of sand, gravel, and boulders are abundant where ice-contact rings are associated with hummocky outwash plains and other glaciofluvial sediments. Contorted beds of lake clay and silt are not uncommon in the moraine-lake plateau environment and are the rule where ice-contact rings formed where lake sediments were deposited on stagnant ice. Where ice-contact rings appear in other lake basins, they are covered with thin lake deposits that appear to have been deposited after the ice-contact rings were formed.

Surfaces of ice-contact rings and ridges are usually strewn with cobbles and boulders deposited in their present position at the time the ring or ridge formed. Some cobbles and boulders undoubtedly accumulated as erosional lag deposits. On photographs, the crests of rings and ridges appear lighter in color than adjacent low areas, in part because of these lag deposits and for other reasons.

To avoid repetitious descriptions, later discussions will largely be concerned with the various origins of ice-contact rings and ridges and their association with other landforms in a hope to shed light on their specific and diverse environments of formation.

Previous Work

The general terms "ice-contact rings" and "ice-contact ridges" and their probable origin were proposed by Parizek (1961, p. 64–94). These terms are applied to landforms referred to by different names by other workers, including "prairie mounds" of Gravenor (1955, p. 475–476); the circular, elliptical, oval, and irregularly shaped "closed disintegration ridges" and "linear disintegration ridges" of Gravenor and Kupsch (1959, p. 53–54); "rim ridges" of Hoppe (1952, p. 5); long, sinuous or elliptical "ice-block ridges" of Deane (1950, p. 14–15); "terrace ridges" and "moraine ridges" of Stalker (1960, p. 8); "circular and linear disintegration ridges" of Bluemle and others (1967); linear "ice-walled gravel trains" of Winters (1963, p. 44–45); "gravel ridges and linear till disintegration ridges" of Bluemle (1965, p. 27, 36); "dump-ridges" of Pettyjohn (1967); rims around "saucer-shaped perched lacustrine plains" of Winters (1963, p. 40–43); rims around "ice-contact lacustrine topography" of Kume and Hansen (1965, p. 22); rims around "elevated lake plains" of Bluemle and others (1967, p. 17); rims around "ice-restricted lake plains" of Hansen (1967, p. 63); rims around "ice-walled lake plains" of Clayton (1967, p. 34); and "doughnuts" of Clayton (1967, p. 32). In addition to the above, the terms "ice-contact rings" and "ice-contact ridges" may be applied to (1) crevasse fillings described by Christiansen (1956, p. 15, 1959, p. 16) and Ellwood (1961, p. 50–73), (2) "ice-crack moraine" described by Sproule (1939, p. 104), and (3) the "ice-block ridges" of Deane (1950). They do not include flutes and grooves or drumlinoidal features or washboard moraines; however, if Clayton (1967, editorial comments) is correct in assuming that many washboard moraines are probably disintegration ridges, they might well be included as ice-contact ridges.

Eskers, crevasse fillings, "ice-crack moraine," and "ice-block ridges" may be included in these two more general groups of landforms termed "ice-contact rings and ridges" for mapping purposes, as they are all of ice-contact origin and more specifically appear to have formed in a stagnant-ice environment largely through glaciofluvial, ablational, and/or ice-press processes. The origins of these latter features have been explained adequately by other workers, however, and they will not be discussed further.

Scope of Investigation

From the variety of terms listed above it should be apparent that descriptive and genetic terms have been coined at will by glacial geologists observing similar landforms in various areas in the Prairie Provinces of Canada and glaciated High Plains states of the United States. It is not my intent to revise the usage of these terms or establish priorities based on first usages in time but rather to propose two terms that encompass all of the various landforms noted above and are both descriptive and genetic. A subclassification of the broader categories "ice-contact rings" and "ice-contact ridges" will be obvious and may very well include some of the terms listed above that are already established in the literature.

A more valuable intent is to present an explanation of the probable origin of various types of ice-contact rings and ridges that have been recognized. The solution to various problems of glacial history can be inferred by the recognition of the presence of ice-contact rings and ridges and an understanding of their origin. In addition, the origin of landforms and deposits with which ice-contact rings and ridges are associated should be better understood as well.

Method of Study

Ice-contact rings and ridges were investigated as part of a glacial geology study in the Willow Bunch Lake area of south-central Saskatchewan (Fig. 1). Aerial photographs at a scale of 4 inches per mile were examined with a stereoscope for the entire area. Ice-contact rings and ridges were grouped according to morphologic and often lithologic similarity and according to their association with other erosional or depositional landforms. Field traverses were made during the summers of 1959 and 1960 at no greater than 4-mile intervals and commonly along all available roads and creeks in search of exposures in conjunction with this and other glacial and ground-water investigations. Fabric studies were conducted using the orientation of tabular lignite chips contained in till comprising various landforms.

Proposed origins of ring and ridge formation are based on their geomorphic form, the composition, texture, and structure of their deposits, association with other landforms and deposits, and a limited knowledge of the behavior of modern glaciers.

The location system, which locates outcrops and landforms to the nearest 10 acres, is shown in Figure 2. It is the standard system used by various Canadian organizations. When two numbers are used alone (5–8), this refers to features within a specific township and range. Three numbers alone refer to the section, township, and range, and so forth.

ACKNOWLEDGMENTS

The Saskatchewan Research Council is the principal sponsor of this field investigation, which was part of a more general study of the glacial and ground-water geology of the Willow Bunch Lake area, Saskatchewan.

Professor P. R. Schaffer, University of Illinois, gave counseling in the initial preparation of a dissertation submitted to the University of Illinois in 1961, in which the body of this report was included. R. E. Borstmayer and J. D. Chapman served as field assistants during the 1959 and 1960 field seasons, respectively, and aided in field studies and in the transfer of photographic data to field maps.

Vertical aerial photographs were taken by the Royal Canadian Air Force and lent to me by the Canadian Department of Agriculture, University of Saskatchewan.

The initial manuscript was written with the financial support of a National Science Foundation Cooperative Fellowship granted while I was at the University of Illinois.

Dr. E. A. Christiansen, Saskatchewan Research Council, lent encouragement to the study and offered editorial suggestions.

ICE-CONTACT RINGS

Rim-Ringed Moraine-Lake Plateaus

The term "rimmed moraine plateau" (Hoppe, 1952, p. 5, 1957, p. 8, 1959, 1963) is applied to undulating to nearly flat areas with raised edges in hummocky ground moraine (Pls. 1–4, 7, and Figs. 3d, 12c). These features have

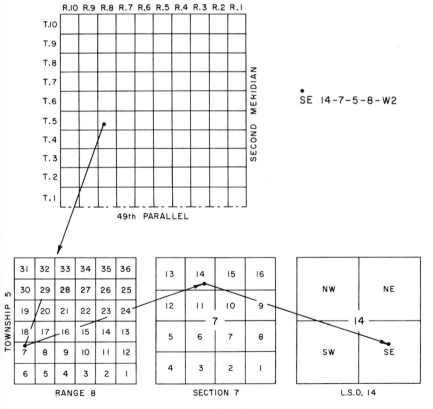

Figure 2. Outcrop location system.

RIM-RINGED MORAINE-LAKE PLATEAU OF HIGH RELIEF

The "smooth" surfaces are covered with glacial lake clay and silt less than 1 to approximately 30 feet thick. Rim rings (R) are well-developed for each moraine-lake plateau. Smaller moraine-lake plateaus (MLP) are present to the east. Elsewhere these grade into single hummocks several hundred feet across. Hummocky ground moraine (HGM) of stagnant-ice origin surrounds the moraine-lake plateaus. A partly buried melt-water channel (BC), developed as a stream trench, appears to the northeast.

Poorly developed ice-contact ridges (ICR) appear along its margin. A late-stage glacial lake (L) developed to the southeast. The lake occupied a topographic low both on top of stagnant ice and, briefly, after ice retreat. The shoreline for this lake is less distinct. It grades into hummocky ground moraine containing contorted superglacial lake deposits (9 and 10–23 and 24).

CLOSE-UP OF RIM-RINGED MORAINE-LAKE PLATEAU OF HIGH RELIEF

Moraine-lake plateaus (MLP) are characterized by "smooth" surfaces underlain by glacial-lake sediments. Multiple ice-contact ridges (MICR) are well developed. They mark ice-floored and ice-walled drainage ways that interconnected moraine-lake plateaus. They lead to a partly buried melt-water channel beyond the area. Rim rings (R) are highly irregular as are rimmed kettles comprising the hummocky ground moraine (HGM), (4, 5, 6, 7, 8–10).

PARIZEK, PLATE 3
Geological Society of America Special Paper 123

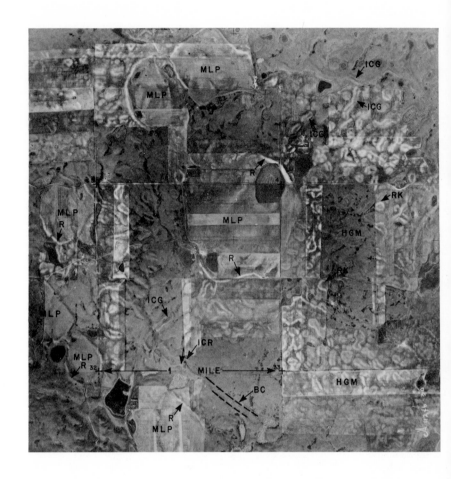

RIM-RINGED MORAINE-LAKE PLATEAUS OF LOW RELIEF

Eight separate moraine-lake plateaus are shown (MLP). Raised rims are well-developed for each (R). Note that moraine-lake plateaus grade into rimmed knobs and kettles (RK), dimpled knobs, and other landforms with which they are genetically related. The hummocky ground moraine (HGM) is composed largely of till of ablation and superglacial origin. Glacial-lake clay caps all moraine-lake plateaus. Beadlike ice-contact rings (ICG) and ridges (ICR) interconnect moraine-lake plateaus. These mark former ice-walled valleys. Ice-contact ridges marginal to a small, partly buried channel (BC) give rise to beadlike ice-contact rings and are interconnected with moraine-lake plateaus (3, 4, 5–4–20; 32, 33, 3, 4–3–20).

been called moraine plateaus and dead-ice plateaus by Stalker (1960, p. 5–11). Stalker called their raised edges "rim ridges," which is descriptive. Ridges encircling the moraine plateaus at lower levels than rim ridges he called "terrace ridges." More recently, Winters (1963, p. 40–43) referred to these same landforms as "saucer-shaped perched lacustrine plains," Kume and Hansen (1965, p. 22) called them "ice-contact lacustrine topography," Bluemle and others (1967, p. 17) "elevated lake plains," Hansen (1967, p. 63), "ice-restricted lake plains," and Clayton (1967, p. 34) "ice-walled lake plains."

"Rim ridges" and "terrace ridges" of moraine-lake plateaus and moraine plateaus are referred to as ice-contact rings in this paper, and to be more specific are called rim rings and terrace rings. These encircling, terracelike, often raised features are genetically related to moraine-lake plateaus and moraine plateaus and hence they are discussed together.

Surface Expression and Composition. The upland surfaces of nearly all large and most small moraine plateaus in the Willow Bunch Lake area are covered with lacustrine silt and clay that range from 1 to 30 feet or more in thickness. Where lacustrine sediments are present on moraine plateaus, they were mapped as moraine-lake plateaus. Where till is present, they are mapped as moraine plateaus. Lake sediments give the upland surfaces of moraine-lake plateaus an even-textured appearance on aerial photographs when compared to adjacent hummocky ground moraine (Pls. 2–4).

Stalker's (1960, p. 6–7) description of moraine plateaus has quite general application to similar landforms noted elsewhere.

The plateaux rise steeply from the adjoining depressions or kettles, commonly at about the angle of repose of the till composing their rims. Along about one quarter of the perimeter of the plateaux, on an average, the margins rise as much as 20 feet above the central parts of the plateaux. This extra rise forms the rim ridge proper. From their summits these ridges slope towards the centres of the plateaux at generally low angles (between about 5 and 15 degrees) and disappear beneath the flat, central parts of the plateaux. . . . The water-deposited sediments of the central parts of these plateaux generally consist of clay, silt, sand and gravel, with included lenses and thin beds of till.

Moraine-lake plateaus are usually from 20 to 150 feet higher in elevation than adjacent areas of hummocky and transitional ground moraine in the Willow Bunch Lake area. In a few instances the surfaces of the plateaus and the surface of the adjacent ground moraine are at approximately the same elevation. The larger moraine-lake plateaus are usually irregular in outline (Pl. 2), whereas the smaller ones are often elliptical to circular to polygonal in plan view (Pl. 4). Moraine-lake plateaus range from a few acres to more than 5 square miles in area in the Willow Bunch Lake area. The largest one reported by Stalker (1960, p. 6) in Alberta is 12 square miles. Their surfaces are often gently to moderately undulating and may be pitted by irregular kettle holes.

Some of the smaller moraine-lake plateaus are moundlike and others are pitted with kettle holes (Pl. 3). The highest point of plateau surfaces need not be centrally located.

All gradations exist in the size of moraine-lake plateaus within a given region. Aerial photographs along common flight lines show dimpled knob topography grading into rimmed-kettle topography that in turn grades into fields of miniature to large-scale moraine-lake plateaus. These relations suggest that these landforms have had somewhat similar origins that varied only in detail.

The relatively flat surfaces of moraine-lake plateaus appear to be related to: (1) initial configuration of the subglacial upland surface, (2) the deglaciation process, (3) presence of lake sediments that tend to fill in depressions, and (4) processes of erosion and deposition that leveled till masses as they were inverted from superglacial to moraine-lake plateau environments in the presence of flowing and ponded melt water.

"Rim ridges" (ice-contact rings or rim rings) are best seen on aerial photographs (Pls. 2, 3, 4). They are irregular to even, sharp-crested ridges that encircle the margins of moraine plateaus and moraine-lake plateaus. They may stand in relief from less than 5 to more than 50 feet above the upland surfaces of plateaus, be flush with their surfaces, or be below them as is commonly the case for plateaus that have somewhat irregular upland surfaces. Rim rings may be breached or obstructed by kettle holes or they may merge with rimmed kettle topography and become obscured (Pl. 4; see centrally located moraine-lake plateau).

Most frequently the rings are one in number, but multiple rings are not uncommon. They may range from less than 100 to nearly 250 feet wide at their base.

These features are called ice-contact rings because they tend to encircle plateaus partly or completely.

"Terrace ridges" or terrace rings may appear along the lower flanks of plateaus at lower elevations than the rim rings. These may be one or more in number. Terrace ridges or rings encircle moraine plateaus. They are not common in the Willow Bunch Lake area.

Rim rings and terrace rings conform to the outline of moraine plateaus and moraine-lake plateaus, indicating that they had a common origin and are genetically related to the plateaus. They may be circular, elliptical, irregular to crenulated. Rim rings may merge with ice-contact ridges that connect one or more moraine-lake plateaus or lead up to open or partly buried melt-water channels.

Most rim-ring and terrace-ring exposures showed that rims and terraces are composed largely of till. Some contained minor amounts of sand and gravel, and in a few instances segments of rim rings were composed entirely of sand and gravel (17 and 18-3-17). Where this is the case, lake-silt deposits predominate over lake-clay deposits on the surface of the moraine-lake plateaus. Till surfaces of rimmed rings and terrace rings are usually strewn with cobbles and boulders. Occasionally, they are capped with thin lake deposits or are composed entirely of contorted laminated silt and clay deposits. Individual moraine-lake plateaus may be interconnected by ice-contact ridges (2-18), (Pls. 3, 4). These connecting ice-contact ridges are composed primarily of poorly sorted sand and gravel with lesser amounts of till.

Lacustrine sediments on the surfaces of moraine-lake plateaus overlie till in all exposures I observed. Localized uplands composed of till were observed to protrude through the lacustrine sediments at a number of localities. Lag cobbles and boulders were common above till in these settings.

Origin of Rim-Ringed Moraine-Lake Plateaus. The origin of this landform has been a subject of controversy. Hoppe (1952), Stalker (1960), and others suggest that the smaller moraine plateaus were formed at the base of a stagnant glacier in dead-ice hollows and that till was pressed from beneath the ice to form rim ridges and lower terrace ridges that have pronounced fabrics. Stalker (1960, p. 18) describes the ice-press process as follows.

During deglaciation crevasses and holes of various types were present in the base of the ice. At the same time the sub-ice material, generally till but in places bedrock or other material, was either not frozen or only partly frozen. It also contained much water, commonly being completely saturated, and thus was in a highly plastic or a fluid condition. The weight of ice on this plastic material pressed it towards the crevasses and holes, increasing the amount of material there while decreasing the amount beneath the ice. In the large holes most of the pressed material came to rest around their margins and only a little near the centre. When the ice finally melted the material that had been pressed into the crevasses and holes stood as ridges, whereas the places from which material had been pressed were low and commonly formed troughs and dead-ice hollows alongside the ice-pressed ridges.

Ellwood (1961, p. 38) favored a "let-down" hypothesis for the formation of moraine plateaus and reasoned that the layer of lacustrine sediments that cap moraine-lake plateaus in Alberta, Canada, "probably insulated the underlying ice over a wider area and promoted more regular downwasting of the ice." He believed that the gently undulating till surface that occurs in knob-and-kettle areas could be produced in this manner.

Hoppe (1952) objected to such a process of slow letdown of superglacial moraine as favored by Ellwood (1961) and other workers. Hoppe felt that rim ridges would not have been produced by this process, but rather flat moraine plateaus would have been formed.

Hoppe (1957, p. 8) describes the best example of irregular moraine plateaus comprising "hummocky moraine landscapes" (moränbacklandskap in

Swedish) or "dead-ice moraine" (dödismorän in Swedish) as irregular masses with hollows between them often some tens of meters deep, and, especially characteristic, each plateau has a rim ridge (likantryggar) around its edge and often other terrace ridges on the slopes down toward the hollows. Hoppe refers to this landscape as of "Veiki" type after a typical locality some 10 km north of Gällivare. His evidence for a subglacial origin of Veiki moraine is the fact that stones in the rim and terrace ridges are found to have preferred orientations transverse to the ridges. Hoppe (1952) states that "this orientation is so regular and so repeatedly observed that it must be regarded as a decisive argument against the Tanner theory (1914)" of their formation by an ablation mechanism. Hoppe describes stone orientations on the plateaus as quite irregular by contrast to the rims and ridges. Hoppe (1957) further states that in many regions the Veiki moraine alternates with drumlin areas, which suggests that drumlins and moraine plateaus were formed subglacially. Of special note is the fact that a region has been found in which drumlins are superimposed on moraine plateaus.

For this locality the following manner of formation is presumed: (1) an uneven deposition of moraine, whereby plateaus and hollows were formed, (2) a formation of drumlins on the surface of the plateaus, (3) a squeezing of water-soaked moraine from the hollows against the plateaus (in this manner the rim ridges were built up). The ice was still moving during (1) and (2), dynamically dead during (3). The intimate connection of the genesis of the plateaus, drumlins and rim ridges is manifested by their practically identical grain size curves. (Hoppe, 1959, p. 312)

Hoppe concludes (1959, p. 205) that Veiki moraine in northern Sweden is unlike similar moraine landscapes in Alberta in that varved sediments cover moraine plateau uplands in Alberta and a core of folded or even overturned bedrock may contribute to the plateaus and ridges. He states that "it is therefore doubtful that the hummocky moraine of Alberta has anything in common genetically with the Veiki moraine." A similar conclusion must be drawn for the Willow Bunch Lake moraine-lake plateaus because they are similar to those that Hoppe and I observed in Alberta. Hoppe visited me in the Willow Bunch Lake area in the summer of 1959 shortly after visiting Alberta. The only significant difference between Alberta and Willow Bunch Lake moraine-lake plateaus is the fact that contorted bedrock is not known to contribute to the upper 5 to 30 feet of moraine-lake plateau sediments in my area.

It is interesting to note that varved sediments are not inconsistent with sedimentation expected in dead-ice hollows above moraine-lake plateaus in view of Hoppe's description (1963) of unique sediment plateaus in the Pälkåive area of northern Sweden.

Stalker (1960, p. 19) recognized the shortcomings of the ice-press theory as advanced by Hoppe because it did not account for the manner in which large moraine plateaus a mile or more in diameter could form at the base of a glacier.

Figure 3 presents an alternate explanation for the origin of moraine-lake plateaus that accounts for the presence of "rim ridges" (rim rings) and the "terrace ridge" (terrace rings) of Stalker (1960), variations in size of moraine-lake plateaus, generally flat upland surface, and the presence of surficial lake clays. This explanation was proposed by Parizek (1961, p. 73–75) and has since been supported by Clayton and Freers (1967, Figs. R-13, R-19) and Clayton and Cherry (1967). As shown in Figure 3, ice loaded with ablation debris has thinned and stagnated. The insulating superglacial debris facilitates differential melting of ice that is accentuated along joints and crevasses above till or bedrock highs. These superglacial depressions foster the development of superglacial lakes. As downwasting continues, pre-existing till or bedrock highs appear as nunataks at the surface of the ice (Fig. 3b). Melting is more rapid around the nunataks, and melt water is ponded above the high against the surrounding ice. Superglacial debris slumps from the downwasting surface of the ice and gives rise to deposits that contribute to the plateau surface and to rim rings and terrace rings at successively lower levels as remnant blocks of ice progressively waste in place. Rings composed of till indicate lack of subsequent sorting. Rings of till, sand, gravel, silt, and clay indicate fluvial and lacustrine processes accompanied their formation. Sediments winnowed from the rings, from adjacent superglacial debris, and from the melting ice are carried into the lake and deposited as lacustrine silt and clay. Where lodgement till is pressed out along the margin of the ice and the moraine plateau, as may be the case at this time, it is added to the rim ring and later to lower terrace rings. As the ice continues to waste, closure for the upland lake is destroyed and the lake is drained by ice-walled valleys or stream trenches, or both, eroded to the base of the glacier. Water contained within these lakes may breach the rings at various places at various stages in development of moraine-lake plateaus. Melt water may drain from one lake-plateau basin to another along ice-walled channels or ice-walled and ice-floored channels until it is discharged into larger, through-going melt-water channels or outwash plains. Superglacial till and other ablation deposits are inverted from their former position on and within the stagnant ice adjacent to the plateau, and terrace ridges or rings are formed at lower levels around the sides of the moraine-lake plateau. Remnant ice blocks finally melt away to produce knob-and-kettle topography, and ponds and sloughs form in the dead-ice hollows (Fig. 3d).

Several observations support this proposed mechanism of rim-ringed moraine-lake–plateau formation. In the Willow Bunch Lake area, most moraine-lake plateaus occur above areas with regional or local bedrock highs as evidenced from contours of the bedrock surface (Parizek, 1964, Pl. II) or in

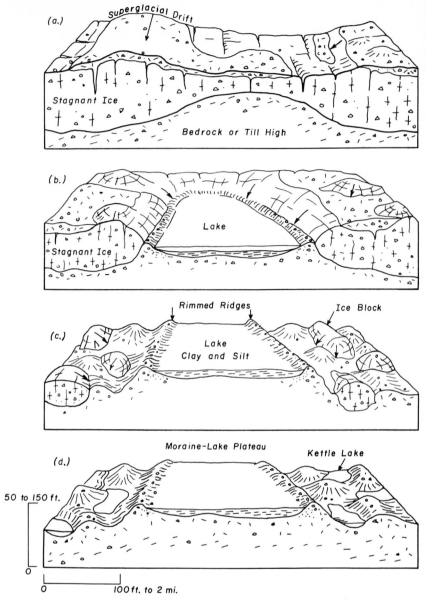

Figure 3. Block diagram showing probable origin of rim-ringed moraine-lake plateaus.
(a) Ice, loaded with ablation debris, has thinned and stagnated. Superglacial debris facilitates
differential melting accentuated along joints and crevasses above till and/or bedrock highs
or where superglacial drift is thin. (b) Depression localizes superglacial lakes where later
nunataks appear. Melting is more rapid around nunataks, and melt water forms moraine-
lake plateaus that receive till and superglacial debris reworked to produce laminated silt
and clay. Superglacial debris slumps from the ice surface to produce rim rings and terrace
rings. These sediments may be deposited as alluvial fan–like deltas or winnowed by wave
action or dumped along the margin of the abandoned lake basins. (c) Isolated blocks of ice
remain as downwasting continues. Terrace rings form between stages (b) and (c). Closure
for lakes is lost. (d) Final topography.

topographically high areas. Local subglacial high areas would be the first to be exposed during downwasting of stagnant ice in the absence of an overabundance of englacial debris that otherwise retards downwasting of underlying ice on both topographic highs and lows. Where moraine-lake plateaus are at nearly the same elevation as adjacent hummocky dead-ice moraine it is possible that moraine-lake plateaus formed in lows in the ice surface that initially contained less superglacial debris and hence melted faster than adjacent ice. These lows later were sites of sediment accumulation that increased their final elevation. By this mechanism subglacial highs need not be called upon.

Orientations of tabular pebbles and cobbles in an exposed rim ring of a moraine-lake plateau show a more nearly random orientation as compared with fabrics obtained from lodgement till in which the effects of ice motion are preserved. These measurements, confined to a partial and isolated exposure of a rim ring and good exposures for hummocky ground moraine support an ablational origin for hummocky ground moraine and rim rings and a mechanism of irregular slumping and dumping of masses of superglacial debris from the surface of stagnant ice.

Closure for these upland lakes had to be provided by ice as uncontorted lake deposits commonly occur at higher elevations than the associated rim rings or the adjacent hummocky ground moraine.

Many moraine-lake plateau basins were interconnected by ice-walled valleys rather than subglacial channels that served as outlets for the lakes. This relation is indicated by the presence of numerous ice-contact ridges that frequently interconnect moraine-lake plateaus and the assumed origin of these features. Subglacial channels and esker ridges, rather than ice-contact ridges, more likely would be associated with moraine-lake plateaus if they were formed in subglacial hollows.

One exposure in a rim ring is especially noteworthy as it clearly illustrates the interfingering of till, till partly washed by melt water, beach sand, and gravel, and lake clays and silts in an ideal strand-line assemblage (Fig. 4). This particular section clearly illustrates the ice-contact origin of rim rings formed adjacent to an open lake basin. Identical relations in some "ice-walled-lake plains" have been reported in North Dakota by Clayton and Cherry (1967, p. 51–52) as further proof that this is not an isolated occurrence.

Hoppe (1952, p. 9) makes a special point of the fact that melt-water channels lead from moraine plateaus, proving that some ice remained on the plateau even at a comparatively late stage. These channels, in his opinion, support the theory that the plateau might be considered a primary formation below the ice. The presence of ice-contact rings and ridges and multiple ridges interconnecting moraine-lake plateaus in my area, on the other hand, support the concept that dead ice afforded closure for the moraine-lake–plateau uplands and that these elevated lake basins were interconnected by way of ice-walled valleys, stream trenches, and perhaps subglacial channels. These glaciofluvial ablational landforms terminate against the moraine-lake plateau flanks within adjacent hummocky ground moraine and occasionally in rim rings or terrace rings. They have never been observed to transect the moraine plateau uplands as would be expected at least rarely if ice completely covered moraine-lake plateaus as postulated by Hoppe.

Hoppe (1952, p. 34–38) describes possible mechanisms for generating basal crevasses in moving glaciers where ice overrides bedrock or till projections. These obstructions should cause compression of the ice in superficial layers and tension in its basal parts capable of fracturing basal ice. He shows how similar fractures should appear parallel to contour lines on the proximal and distal sides of subglacial projections. The crevasses postulated by Hoppe would tend to remain open if formed just prior to mass stagnation and would facilitate downwasting of ice above local obstructions to produce nunataks necesssary to the formation of moraine-lake plateaus as outlined in this report.

Moraine-lake plateaus are especially well-developed northeast of Ormiston and northwest of Ceylon, Saskatchewan, where regional bedrock or drift highs, or both, have been recognized (Pl. 1). These uplands are bordered by re-entrant lowlands on either side that should have localized the more rapidly moving portions of the glacier in a manner postulated by Hoppe. Tension should have developed due to the fact that localized streams of ice moved faster along these lowlands than did the lateral parts overlying topographic highs. It is reasonable to assume that tension crevasses would have been well-developed in this environment. Transverse crevasses also would develop above these subglacial uplands because the glacier was forced to

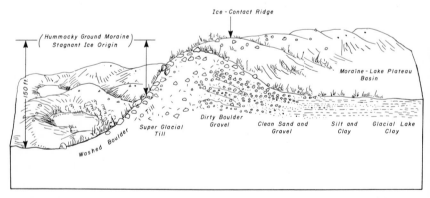

Figure 4. Cross section through a rimmed ridge or rim ring showing transition from ice-contact deposits to lake silts and clays capping a moraine-lake plateau.

override regional bedrock uplands. Crevasses produced by a combination of these processes might facilitate nunatak development even in regions of thick ablation and superglacial drift. Once differential melting favored first superglacial-lake–basin development and later lake development above the emerging nunataks the process of moraine-lake–plateau development was accentuated.

Outlets for melt water contained in moraine-lake basins appears to have been controlled by the elevation of melt-water channel floors eroded across transverse ridges interconnecting these lake basins, across bedrock highs, end moraines developed on dead ice, and irregular masses of dead-ice moraine. Containment of melt water or at least restricted discharge of melt water derived from thin masses of stagnant ice appear to have favored the development of moraine-lake plateaus. Commonly, moraine-lake plateaus are well-developed in the Willow Bunch Lake area where stagnant ice was downwasting in the presence of abundant superglacial drift, where the main mass of the glacier was retreating down regional slope, and where well-developed hummocky end moraines (Maxwellton and Ardill End Moraines (Pl. 1)) and high-level outwash plains (Ormiston region) controlled the level of the free discharge of melt water. Elsewhere, as near Crane Valley (Pl. 1), proglacial lakes may have served as a temporary base level that controlled local moraine-lake–plateau development. A more regional base-level control for moraine-lake–plateau development was suggested by Clayton and Cherry (1967, p. 48).

Clayton and Cherry (1967, p. 48–50) have contributed further to the understanding of the origin of moraine-lake plateaus. They point out that "ice-walled lakes" may be distinguished from superglacial lakes by having been bottomed on solid ground, not on ice, and were surrounded by stagnant ice. They distinguish two end members, coalesced and uncoalesced basins. Their evidence that sediment was deposited around ice blocks in coalesced basins is as follows:

(a) The high relief topography consists largely of lake sediment resting on a relatively flat surface of till. (b) The areas that were presumably between the ice blocks are relatively flat plateaus underlain by uncollapsed lake sediment. (c) The plateaus rise no higher than 1675 and 1650 feet; they do not exceed the maximum water level of 1700 feet, determined by the elevation of the outlet of associated Lake Saltcoats (Christiansen, 1960, p. 16). (d) The areas which were presumably occupied by the ice blocks contain no hummocky collapse topography or superglacial lake sediment, indicating that the ice-blocks extended above water level. . . . (e) The plateaus grade into areas of hummocky collapse topography at several locations in the Yorkton (Saskatchewan) area. Presumably the ice blocks extended below the lake in places, resulting in superglacial deposition. . . .

Uncoalesced "ice-walled lake basins" by contrast occupied single, separate ice basins that contained lakes 1 or 2 miles wide (Clayton and Cherry,

1967, p. 49). They further distinguish two end members for "uncoalesced ice-walled lake basins": (1) lakes formed in a stable environment with thick superglacial till on the surrounding stagnant glacier, and (2) lakes formed in an unstable environment with thin superglacial till on the surrounding ice. The following remarks by Clayton and Cherry (1967, p. 49–50) were appropriate.

The superglacial till in the unstable environment was thin (a few tens of feet or less), and ice melted rapidly. (Slopes in the resulting dead-ice moraine are gentle, 2 to 7 degrees, and the relief is low.) As a result, the topography on the glacier underwent continuous change, and mass movement of the superglacial till was very common. Much superglacial till slumped or flowed into the lakes, and cold silt-laden superglacial rivers carried much sand and gravel into the lake margins. Waves sorted and redistributed the mudflow and slump material, carrying the fines out into the lake, leaving the sand and gravel fraction on the beach along with the sand and gravel brought there by the rivers. Meltwater frequently moved through the lakes with a velocity sufficient to prevent the deposition of the finest suspended particles. Therefore, the lake sediments are typically much coarser than the stable-lake sediments. In the Ross area, the sediment is commonly clayey silt in the center of the lake plains, and gradually coarsens to sandy gravel at the margins.

Lake plains formed in the unstable environment, such as those near Ross, are concave upward. This is a result of the great amount of coarser material deposited near shore and the greater compaction of the finer mid-lake deposits.

Unstable-environment lake plains commonly have either till or gravel rims—till where the superglacial till slumped off the ice into the lake margin without being reworked by the waves, and gravel where it was reworked by the waves.

The plains of lakes in the unstable environment tend to be round in outline because of the greater shore activity than in the lakes in the stable superglacial environment; shore activity tended to remove any irregularities projecting into the lakes.

. . . In the stable superglacial environment, the superglacial till was so thick (several tens of feet) that the ice melted very slowly. (Slopes in the resulting dead-ice moraine are presently steep, 7 to 20 degrees, and local relief is high; . . .) As a result, the topography of the stagnant glacier changed very slowly, and mass movement of the superglacial till was infrequent. Superglacial rivers were relatively tranquil and carried little silt; only the material that settled out very slowly reached the lakes. Therefore, the sediments that accumulated in these lakes are very fine grained (clay and silty clay), such as in the lake plains near Tagus.

Lake plains formed in the stable environment, such as those near Tagus, are commonly convex upward because there was relatively uniform deposition of clay over the entire lake bottom and the margins rested on ice; the margins collapsed when the ice melted.

Stable-environment lake plains do not have rims because there was little slumping of material off the stable ice walls.

Sediments deposited in the stable-environment lakes are commonly much thicker than those deposited in the unstable-environment lakes. Lake plains of the stable environment are perched above the surrounding dead-ice moraine, whereas the plains of the unstable environment are commonly in low areas, but are above the bottoms of depressions immediately adjacent to them. Apparently the lakes in a stable environment lasted for a much greater length of time, allowing more lake sediment to accumulate than in the short-lived lakes of the unstable environment.

Although an adequate sampling has not been made, aquatic fossils apparently are more abundant in lake sediments deposited in stable-environment lakes because the water was warmer, more silt-free, and there was more time for the establishment of plant and animal populations than in the unstable environment.

These ideas represent a valuable extension of origin of moraine-lake plateaus that I proposed in my 1961 dissertation.

Rimmed Kettles

"Prairie mounds" (Gravenor, 1955, p. 475), "rimmed kettles" (Christiansen, 1956, p. 11), "closed disintegration ridges" (Gravenor and Kupsch, 1959, p. 52), "doughnuts" (Clayton, 1967, p. 32), "humpies," and similar names have been applied to low-relief landforms that are nearly circular in plain view and in most cases have central depressions that may be round or somewhat irregular (Pls. 2, 3, 4, 9). These features may resemble doughnutlike features, dimpled knobs, crenulated knobs, and knobs resembling barley grains. Commonly in the Willow Bunch Lake area all variations of these landforms may be noted on aerial photographs along a particular flight line where normally one landform grades into another in a transitional manner.

Gravenor and Kupsch (1959, p. 52–53) describe "closed disintegration ridges" as follows.

. . . Perfectly circular ridges range in height from a few feet to 20 feet and in diameter from 20 to 1,000 feet. The ridge, which surrounds a central depression, is commonly referred to as a "rim." It is generally unbroken and of the same height around the depression. Concentric ridges have been noted, and in places a low circular knoll takes the place of the more common central depression. The circular ridges are similar to one type of "prairie mounds" described by Gravenor (1955), but in "prairie mounds" the base of the central depression lies well above the general ground level, whereas the base of the central depression in circular ridges lies at or below that level. Intermediate forms, however, exist. In general, the low circular ridges surround a depression that is shallow, but the floor is somewhat below the general level of the till plain.

Locally the circular ridges occur with ridges that are oval or irregular in plan. . . . They surround depressions which may vary in shape from circular to irregular. All such ridges are referred to as *closed ridges* by the authors, even though they may show some minor breaks resulting from irregular deposition or later erosion.

. . . Most authors seem to agree that "closed ridges" resulted from stagnant ice that separated into dead-ice blocks, even though they do not agree as to the mechanics of deposition. . . .

The landforms described above by Gravenor (1955) and Gravenor and Kupsch (1959) are similar to those observed in the Willow Bunch Lake area and elsewhere with the exception that in my area the relief may be more pronounced, from 5 to nearly 75 feet between the ring crest and adjacent low areas. Higher relief is typical where knob-and-kettle topography is well-developed. In these areas, however, the rimmed-kettle outline normally gives rise to irregularly shaped rimmed kettles (Pl. 3) or knob-and-kettle topography.

Rimmed kettles may be composed of till, sand, gravel, laminated silt and clay, or a combination of two or more of these. Irregular masses of glaciofluvial and glaciolacustrine deposits are more abundant where these landforms were derived from stagnant ice covered with lacustrine sediments and outwash. Bedding usually is deformed.

A further description of this widespread and rather well-known landform is not warranted here.

Probable Origin. Henderson (1952) favored a mechanism of mound formation under a periglacial environment where the ground was permanently frozen and where the centers of these mounds were raised by the growth of polygonal ice wedges. More recently, Bik (1967) proposed a mechanism of collapsed-pingo formation combined with subsurface displacement of plastic material to explain the origin of all prairie mounds. He concludes (p. 87) that "inasmuch as the prairie mounds occurring on till are identical in shape and form to those occurring on proglacial lacustrine deposits, the landform cannot have a subglacial or supraglacial genesis, but is either of subaquatic or subaerial origin. Furthermore, as the parent form in the case under discussion is made up of basal till, the genesis of the form is probably the result of deformation of an already existing deposit . . . , rather than of concurrence of deposition and deformation."

A letdown hypothesis was favored by Gravenor (1955, p. 476–477) for the origin of prairie mounds. Gravenor and Kupsch (1959) allowed for a combination of several depositional processes to explain the origin of dead-ice landforms, including prairie mounds. For this reason they (p. 60) preferred that the term "disintegration features" rather than "ablation moraine" be applied to a host of landforms originating from stagnant ice. By this designation they allowed for a combination of ablation and basal squeezing to account for the origin of disintegration features, including landforms referred to as ice-contact rings in this paper.

Stalker (1960) favored a largely basal ice-pressed drift origin for similar landforms caused by the intrusion of plastic basal till into subglacial cavities developed beneath stagnating ice masses.

More recently, Clayton (1967, p. 30–32) showed diagrammatically three stages in the formation of circular disintegration ridges or "doughnuts" (ice-contact rings) that is in agreement with Gravenor's proposed origin (1955) for prairie mounds. The mechanism includes (1) sliding or flowing of superglacial till or lake sediment into a sinkhole in the stagnant ice, (2) inversion of topography as a result of the insulating effect of the drift in the bottom of the sinkhole, followed by mass movement of this drift away from the center and down the sides of the buried ice core, and (3) melting of the ice core.

This latter process is preferred for the formation of most ice-contact rings of the rimmed-kettle type in the Willow Bunch Lake area. Where superglacial lake sediments and glacial outwash are known to have been present on stagnant ice in my map area similar rimmed kettle landforms resulted that are composed almost entirely of stratified drift. Collapse, letdown, and inversion processes account for their topographic form and contorted bedding, the lack of uniformity of till fabrics observed, and their association with related dead-ice landforms. If an ice-press mechanism operated as well, there is no evidence that this process dominated over ablational processes.

ICE-CONTACT RIDGES

Ice-Contact Ridges Marginal to Melt-Water Channels

Gravenor and Kupsch (1959, p. 55) describe two types of melt-water channels that are common in areas that were covered by stagnant ice. There are (1) those that are partly or almost completely filled with till and/or other drift deposits, which are recognizable on aerial photographs by a chain of kettles in the bottom of the valley and by gullied, abrupt breaks in slope that mark their former margin with the adjacent upland surface, and (2) those that are not filled with till but are broad, open troughs with some sand and gravel on their floors.

Three types of partly filled melt-water channels may be distinguished. (1) There are those that were formed earlier, then overridden by glacial ice and only partly buried during the glacier's retreat. This is particularly common where melt-water channels formed marginal to the ice front of an active glacier. Stagnant-ice blocks often persisted in these channels after ice abandoned the adjacent uplands. (2) There are those partly filled channels that formed in an ice-walled environment but where the floor of the channel was eroded largely below the base of the glacier into lodgement till and in some places bedrock. Segments of some of these channels were, in part, subglacial channels that were later partly filled by ablational drift after they were abandoned by melt water but before complete ice retreat. (3) Other channels were ice-walled but not ice-roofed. In these, superglacial drift partly filled the channels after they were largely abandoned by melt water but before complete ice retreat.

Open melt-water channels were formed in both ice-walled and proglacial environments. Where open channels were ice-walled, either ablational drift was not available to partly fill the channel or melt water continued to occupy the channel and erode away ablational drift as it was dumped from the stagnant ice surface until adjacent uplands were deglaciated.

All but proglacial melt-water channels may have ice-contact ridges along the brim of their upland margins as an indication that the channels formed in an ice-walled environment (Pl. 1). In some cases channels formerly eroded in an ice-walled environment were later used by proglacial melt water during a later ice advance and retreat. Melt-water channels with a double history may have ice-contact ridges along their margins as well. However, the ice-contact ridges were formed during the disintegration of earlier masses of ice.

General Form and Composition. Normally these ice-contact ridges are sharp-crested ridges that may vary from a few hundred feet to 5 miles or more in length. They may have irregular crests when seen in longitudinal profile or their crests may be discontinuous. They may range from about 5 feet to nearly 35 feet in height and from 50 to 150 feet in width at their base. Single ridges are most common, but rarely composite subparallel to parallel ridges have been observed along either or both channel brims. Most commonly two single ridges are present, one on each side of the channel brim.

These ice-contact ridges are composed of ablation till, sand, and gravel with little if any bedding preserved, or with deformed beddings, and they may contain contorted beds of laminated silt and clay. Typically they have cobble and boulder lags on their crests. Any and all combinations of these deposits may occur together in a given exposure. In this assemblage contacts between sedimentary units may be irregular, having been deformed from their original position.

Probable Origin. In all cases described, ice-contact ridges bordering melt-water channels are interpreted to have formed in a stagnant-glacier environment whereby supraglacial and englacial drift was let down along the margins of dead-ice blocks parallel to active and abandoned melt-water channels. An ice-press mechanism may have contributed subglacial debris as well but probably to a lesser extent.

An explanation involving an ablation process is preferred. It is reasoned that if an ice-press mechanism predominated subglacial drift would have been squeezed out into the melt-water channel rather than upward to form a delicate ridge at the very brim of the channel. Also, ice-contact ridges in this setting frequently merge with "circular disintegration ridges" or "doughnuts" that, according to Gravenor (1955, p. 476–478), Gravenor and Kupsch (1959), and Clayton (1967, p. 31–32) formed largely from an inversion process whereby superglacial drift was systematically "let down" to its present position and form. Ice-marginal ridges and circular disintegration ridges are believed to be genetically related when they occur together in this association.

Ice-Contact Ridges Adjacent to Channels Occupied by Stagnant Ice. In some cases, stagnant-ice blocks appear to have occupied the channel at the time of ridge formation, and, in other cases, ice appears to have covered the uplands along the margins of the channels. Just south of the Maxwellton End Moraine (Pls. 1, 5) and west of Maxwellton (9–27, 28, 29, and 30), numerous partly filled, roughly parallel melt-water channels occur more or less in side-hill position or at right angles to the regional topographic slope. It is postulated that the ice that deposited the Maxwellton End Moraine (Parizek, 1964) advanced southwestward across the Missouri Coteau Upland (Pl. 1) and the Old Wives Lake Plain and disrupted the proglacial drainage system that continually formed in ice-marginal position until the ice overrode the drainage divide (Pl. 5). Melt-water channels (K, L, and M in Fig. 5a) formed along the margin of the ice and then were overridden as ice continued to advance up regional slope. Witkind (1959, p. 20–21) describes similar ice-marginal channels in the Smoke Creek–Medicine Lake–Grenora area of Montana and North Dakota to the south, but he did not indicate whether or not ice-contact ridges were present along their rim. Rather he recognized the fact that ice-marginal channels may form during halts of an advancing ice sheet and later be overridden by a renewed ice advance moving up regional slope. Till-covered, outwash-free channels are believed by Witkind to have formed by this process, particularly channels that are now reflected by elongate undrained depressions occupied by small lakes. Clayton (1967, editorial comments) indicates that partly filled and open melt-water channels with ice-contact ridges along their rims also are common on the Missouri Coteau in North Dakota.

With melting of the ice to the condition shown in Figure 5b, ablation and superglacial till gave rise to knobs and kettles with low relief and other landforms characteristic of areas of transitional and hummocky ground moraine. Remnant blocks of stagnant ice, 50 to 150 feet thick and 0.25 to 0.5 mile wide still occupied the melt-water channel shown in a close-up view in Figure 5b. Ablation and superglacial drift sloughed off either side of the ice block that occupied the channel and gave rise to two parallel ice-contact ridges. With further melting, the remainder of the superglacial and ablational debris was inverted from its position on and in the ice and hummocks of till were deposited along the channel walls and on the channel floor (Fig. 5c). Chains of lakes or sloughs now fill depressions that were occupied by the last isolated blocks of ice. Much of the region on either side of the Maxwellton End Moraine shown on Plate 5 was later covered by thin lake sediments that have subdued the features in the area in question.

Some ice-contact ridges may have formed along the margins of partly buried channels that were eroded during previous glacial episodes. After

ICE-CONTACT RIDGES MARGINAL TO
PARTLY BURIED MELT-WATER CHANNELS

Partly buried channels (BC). Ice-contact ridges (ICR) parallel to partly buried channels, parallel to the crest of Maxwellton End Moraine and of irregular nature. Glacial lakes developed immediately south and north of the end moraine during and after deglaciation. Ice-contact ridges parallel to the end moraine developed where stagnant ice persisted buried within end moraine deposits after ice downwasted and retreated downslope to the north. Closure for the northern lake was afforded by the end moraine deposits and enclosed ice masses and the retreating stagnant glacier to the north. The partly buried channels formed earlier in an ice-marginal environment. Melt water channels proxy for the end moraine to the northwest where drift was eroded by melt water (9 and 10–29).

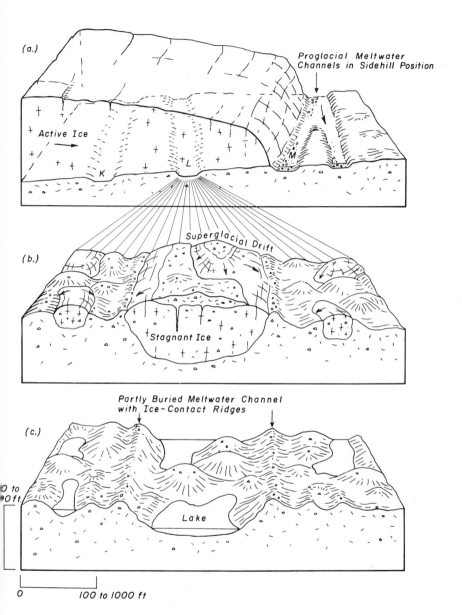

Figure 5. Probable origin of ice-contact ridges along melt-water channels occupied by stagnant ice. (a) Ice advances up regional slope. Proglacial melt water forms ice-marginal channels that later are overridden by the glacier. (b) During retreat, ice thins and stagnates. Isolated blocks of ice occupy melt-water channels. Superglacial drift slumps to produce ice-contact ridges, rimmed kettles, and related landforms. (c) Final topography.

these channels were eroded they were overridden by ice. As the ice thinned during the next retreat, masses stagnated in the channels and ice-contact ridges were deposited as discussed above.

Ice-Contact Ridges Adjacent to Stream Trenches and Subglacial Channels. Figure 6 shows the proposed origin for ice-contact ridges formed along the margins of another type of partly filled melt-water channel. These channels are thought to have been eroded in the ice and into the underlying ground moraine and/or bedrock. Channels of this type are termed "stream trenches" by Gravenor and Bayrock (1956) (*see* Pls. 2, 4). Entire channels or segments of these channels may have formed in a subglacial environment. Others may have had walls sufficiently high and ablation debris sufficiently abundant to account for the knobs of till, contorted sand, gravel, silt, and clay deposits that partly filled these channels after they were largely abandoned by melt water.

Partly filled channels with or without marginal ice-contact ridges are widespread on the Missouri Coteau in Saskatchewan and North Dakota and elsewhere where dead-ice moraines are well-developed.

The courses of partly filled melt-water channels that formed as ice-walled channels or stream trenches usually have straight channel segments and right-angle bends, their tributaries join them at right angles, and channels may breach topographically high areas. Stream trenches may end abruptly against glacial drift or bedrock at the point of their maximum headward development. Channel segments in low areas may lead into constructional ridges of sand and gravel with primary depositional surfaces and structures still preserved. Other segments along the same channel in topographically high areas may be erosional. Stream trenches of ice-walled origin usually are found associated with hummocky ground moraine or transitional ground moraine (13–4–23, 24 and 25–5–22, 17–4–19) and some hummocky end moraines.

The mode of origin of these ice-contact ridges is similar to other ice-contact ridges formed along partly filled channels except that the ice stood on the uplands adjacent to the channel during ridge formation rather than in the channel bottom as previously discussed. It is not always possible to distinguish channels and ridges that were formed in an ice-walled environment from those formed while the ice stood in the channel because both channels may have hummocks of sand, gravel, till, and clay on their floors and ice-contact ridges along their rims. Channels of ice-walled origin usually have courses that are angular and tributary channels that join at nearly right angles. The rectangular drainage pattern is believed to have been inherited from joints and crevasses developed during ice motion and later reflected in the stagnant ice. Channels may be somewhat regularly spaced, parallel to the direction of ice advance, and at right angles. Channel orientations may reflect subtle

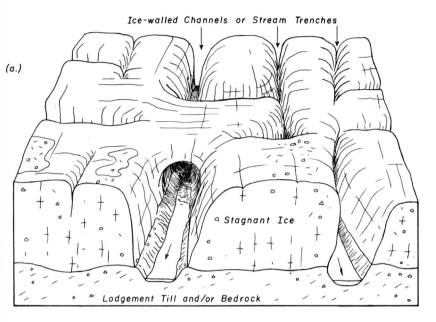

(a.)

Ice-walled Channels or Stream Trenches

Stagnant Ice

Lodgement Till and/or Bedrock

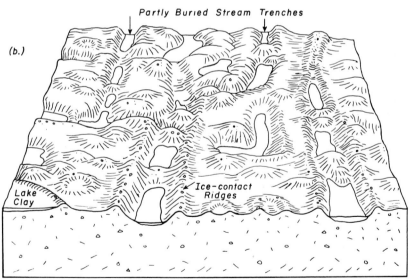

(b.)

Partly Buried Stream Trenches

Lake Clay

Ice-contact Ridges

Figure 6. Probable origin of ice-contact ridges formed along stream trenches. (a) Ice-walled channels and stream trenches develop along crevasses in stagnant ice. Ablation and superglacial debris give rise to ice-contact ridges, partly buried channels, and other landforms of dead-ice origin. (b) Final landforms showing angular bends and tributary channels reflecting primary control by ice. Hummocky channel fill deposits indicate that many channels were abandoned before complete downwasting of adjacent and/or overlying ice.

radiating and circular patterns suggestive of control by joints developed near the snouts. Channels may be superimposed across local bedrock or till in topographically high areas, which further supports the ice-wall origin. Segments of ice-walled channels may be free of ablation deposits, particularly where the adjacent uplands lack strong development of knob-and-kettle topography. Other segments may be traced into hummocky end moraines and still not be completely filled by drift, suggesting that they served as open raceways for melt water at the time the end moraine was being deposited. Abandoned melt-water channels eroded during an earlier episode and subsequently covered by end-moraine debris are usually completely buried. This is true for the partly buried channel that extends from Mitchellton to Montaque Lake, where it is buried beneath the Maxwellton and Ardill End Moraines (Pl. 1). In contrast, channels in the Maxwellton End Moraine southwest of Ceylon (4–21, 4 and 5–22) are only partly buried, although they are within a well-developed hummocky end moraine. The channels west of Ceylon are believed to have formed as ice-walled channels during the deposition of the end moraine

Ice-Contact Ridges Marginal to Open Melt-Water Channels. Ice-contact ridges marginal to open melt-water channels are identical to those that are marginal to partly filled channels described earlier. The main distinction is that they were formed marginal to open ice-walled channels that remained entirely or almost entirely free of ablation debris after the ridges were formed. Open melt-water channels may contain valley-train sand and gravel, more recent stream alluvium and colluvium, or glacial-lake sediments.

Open melt-water channels in one area may be traced into partly buried channels in an immediately adjacent area. Ice-contact ridges may be present in either or both areas, suggesting that although the channels may have slightly different origins, their raised rims had identical origins.

Plate 6 shows an open ice-walled channel with well-developed ice-contact ridges on either side. The meandering pattern of the channel, the presence of the ice-contact ridges along its margin, and other landforms in adjacent areas derived from stagnant ice suggest that this channel was entrenched in a field of stagnant ice. The segment of the channel containing Dryboro Lake (2, 3, 4, 5, 8, 9, 10, and 11–27–27) is shown. The pronounced ice-contact ridges are not clear without the aid of a stereoscope; hence a segment is outlined. Melt water flowed down this channel through a field of stagnant ice at least 12 miles wide.

The sinuous course of this and other melt-water channels probably developed initially on the surface of stagnant ice, and, as the channel was downcut through the ice, the meander pattern was superimposed on the underlying ground moraine and bedrock (Fig. 7). The meanders continued to develop lat-

ICE-CONTACT RIDGES MARGINAL TO AN OPEN MELT-WATER CHANNEL

Meander bends are well shown. Variable elevations of land surface, confined course of channel irrespective of topography, ice-contact ridges marginal to the channel, and associated landforms of stagnant ice origin indicate ice-walled origin for the channel. Hummocky ground moraine (HGM), ice-contact ridges (ICR), channel partly buried to the east (BC) where it was ice-floored (3, 4, 9, 10, 15, 16–9–27).

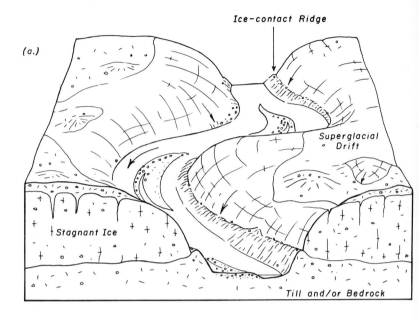

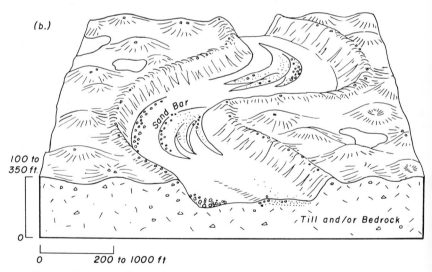

Figure 7. Probable origin of ice-contact ridges marginal to open melt-water channels. (a) Meandering channel developed on stagnant ice. Channel is incised into the underlying drift and bedrock and the meander continues to develop. Superglacial drift cascades from the ice to give rise to ice-contact ridges marginal to the melt-water channel. Features last formed survive erosion. Knob-and-kettle topography, dimpled knobs, rimmed kettles, and similar dead-ice landforms develop adjacent to the channel. (b) Final landscape. Similar ice-walled channels may form along crevasses to produce straight or irregular valley segments with parallel ice-contact ridges.

erally in the ice-walled channel until the volume of flow diminished. Sand and gravel were deposited by the overloaded stream and ultimately the channel ceased to migrate laterally. Prior to this, ablational debris slumped from the ice surface and was swept away by melt water. When the channel ceased cutting laterally, the ice-contact ridges, composed of till, boulders, sand, and gravel, were deposited on ground moraine along the margins of the channel by the inversion mechanism described earlier. Two sinuous ice-contact ridges were formed on either side of the channel in both topographically high and low areas as further proof that the channel was ice-walled. As these ridges formed (Fig. 7), the ice thinned to the extent that ablation and superglacial drift were deposited to form hummocks, rimmed kettles, dimpled knobs, and other landforms of stagnation origin on the adjacent uplands.

The main difference between the type of channel cited above and those shown in Figure 6 is that, for the former, ice must have stood high above relatively narrow channels and even arched over the stream trenches in some areas to supply the debris that has since partly filled them in. For the latter, channels were ice-walled but not overtopped by ice during their late-phase development.

Rings and Ridges Formed in Ice-Floored and Ice-Walled Valleys

Ice-floored valleys formed in active and stagnant ice where melt water flowed down the consequent slope of the glacier's surface, where melt water tunnels developed within the glacier, and where open crevasses were produced in response to the glacier's motion. In the Willow Bunch Lake area, some of these valleys were walled and floored entirely by ice during their early history or they may have given rise to valleys that ultimately extended to the base of the glacier to form ice-walled valleys or stream trenches. Many stream trenches were ice-floored along segments of their courses and others were initiated as ice-floored valleys.

Landforms produced from stagnant ice and localized by ice-floored and ice-walled valleys may be regular or somewhat irregular in their appearance and distribution, depending upon whether they were produced by uncontrolled or controlled glacial disintegration. Gravenor and Kupsch (1959, p. 49) state that

When the forces that operate to break up an ice sheet are equal in all directions, the disintegration may be said to be *uncontrolled,* and the result is a field of round, oval, rudely hexagonal or polygonal features, and a general lack of dominant linear elements. When the ice separated along fractures or other lines of weakness, the disintegration may be said to be *controlled,* and the result is a field of linear or lobate landforms. In places the ice broke along open crevasses or along thrust planes, both of which formed when the ice was still flowing, and the disintegration thus shows *inherited flow control.* . . .

Landforms produced from inversion of channel-fill sediments localized within ice-floored valleys always tend to reflect "controlled disintegration" in that channels were localized by crevasses, and valleys developed on the consequent slope or developed as meandering channels on the ice surface. Two end members are recognized for "controlled disintegration" where valleys were involved. There are those ice-walled and ice-floored valleys that were controlled by systematic crevasses and low areas on the ice surface initially but whose regularity became modified by through-going melt water as ice disintegration progressed. Landforms produced in this valley environment are comprised largely of glaciofluvial sediments. The other end member is represented by landforms whose regularity of distribution reflects systematic planes of weakness within the glacier that were little modified during disintegration. Landforms produced in this setting are composed largely of till, laminated silts and clay, and lesser amounts of glaciofluvial sediments. "Linear disintegration ridges" or ice-contact ridges formed in the latter environment characteristically display systematic patterns as if controlled by systematic joints developed within the glacier. On a regional basis these landforms may define lobate patterns. Individual ridges may be parallel, nearly at right angles, or oblique to the direction of ice movement. Gravenor and Kupsch (1959, p. 53) indicate that one set of prominent ridges commonly lies at 45° to the direction of ice motion.

Landforms produced in the ice-floored and ice-walled environment have been referred to as "ice-crack moraine" by Sproule (1939, p. 104), "crevasse fillings" by Flint (1928, p. 415), "ice block ridges" by Deane (1950, p. 14), "till crevasse fillings" by Gravenor (1955, p. 10), "linear disintegration ridges" by Gravenor and Kupsch (1959), and "ice-contact ridges" by Parizek (1961).

Of particular interest here are those landforms that appear to have formed in an ice-floored crevasse environment above the base of the glacier and that were derived from glaciofluvial, superglacial, and englacial sediments inverted from their former position on the crevasse floor during the deglaciation process. The origin and nature of ice-contact ridges formed by an ice-press and ablational mechanisms have been described by other workers.

Ice-contact ridges derived from ice-walled and ice-floored channels appear as anastomosing or branching ridges that may have one or more crests. Their crests may branch and merge in a somewhat irregular manner or they may give rise to aligned ice-contact rings that resemble links in a chain, in which case they are called chainlike ice-contact rings. Ice-contact ridges and rings of these types are thought to have been formed in the same environment; hence they are discussed together.

Ice-contact ridges and chainlike ice-contact rings often connect two or more moraine-lake plateaus (Pls. 3, 4) or they may be aligned or branching as shown on Plate 7 or display meanders (Fig. 8). They frequently

ICE-CONTACT RIDGES AND CHAINLIKE RIDGES

Note the branching character of chainlike ridges (ICG) and multiple ice-contact ridges (MICR) and the parallel-to-angular channel relations suggesting that joint control localized ice-walled valleys. Ice-contact rings and ridges may join moraine-lake plateaus (MLP) and give rise and merge into dimpled knobs, rimmed knobs, rimmed kettles, and knobs and kettles.

Ice-contact ridges (ICR) border partly buried channels (BC) that merge into chainlike ice-contact rings (ICG) and ridges (13, 14, 18, 22, 23, 24–2–19).

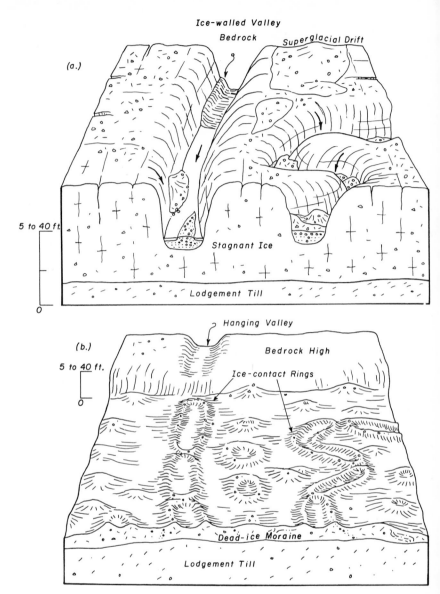

Figure 8. Ice-contact rings and chainlike ice-contact ridges formed in ice-walled valleys. (a) Ice-walled valleys may be joint or crevasse-controlled or they may develop as meandering or irregular-shaped valleys on the surface of stagnant ice. As valley bottoms extend downward, bedrock and drift topographically high areas may be breached by melt water later to produce hanging valleys or wind gaps. Superglacial drift cascades into these valleys and may be reworked by melt water. As alternate routes become available ice-walled valleys are abandoned. (b) After downwasting is complete, ice-contact rings, chainlike ridges, and multiple ridges may result as shown in Figures 9 and 10. Knob-and-kettle topography, dimpled knobs, rimmed kettles, and similar landforms develop in adjacent areas by somewhat similar ablational processes.

occur in low areas of hummocky ground moraine of stagnant-ice origin. They may lead up to and terminate at open or partly buried melt-water channels, melt-water channels that contain marginal ice-contact ridges, or the base of hanging melt-water channels. These features are similar in general form and composition to ice-contact rings and ridges described earlier with the one exception that they tend to contain a greater abundance of glaciofluvial sediments.

Commonly a well-developed ridge may give rise to and merge into a single knob or a field of dimpled knobs, rimmed kettles, crenulated knobs, or similar forms. One can conclude from their association with these landforms and their connection with ice-contact rings marginal to moraine-lake plateaus that all of these landforms were produced in a somewhat similar manner, largely by an ablational process that involved one or more inversions from their former position within or on the surface of stagnant ice.

Figures 8, 9, and 10 show the proposed origins for ice-contact rings that resemble links in a chain and single-crested and multiple-crested ice-contact ridges. Valley-train sand and gravel and/or ablational debris were deposited in ice-walled valleys enlarged from crevasses and joints developed above the base of the glacier and in meandering channels formed in a superglacial environment. In some areas, many ice-walled channels are parallel and regularly spaced and appear to preserve crevasse patterns developed earlier in the active glacier.

Evidence for an ice-walled and ice-floored origin for these features was noted at several places where ice-contact rings were associated with hanging melt-water channels. In Figure 8, ice-contact rings terminated against bedrock and till, topographically high areas breached by melt water. The elevation of these "hanging channels" marks the elevation of the graded channel floor eroded in the surrounding ice. As the ice wasted below the surface of the subglacial till or bedrock highs, the channels were abandoned and wind gaps or hanging melt-water channels were left 30 or more feet above the surface of the adjacent hummocky ground moraine deposited later by the waning ice. Similar landforms have been observed to terminate against hanging outwash deposits 10 to 50 feet higher in elevation than adjacent ground moraine containing ice-contact rings. Glaciofluvial sediments being transported by melt water were abandoned in the ice-walled channels as melt water drained from the field of stagnant ice along alternate lower routes. The ablational and superglacial till that slid into the ice valleys and the glaciofluvial sediments deposited earlier served to insulate the underlying ice. The thicker uninsulated ice walls wasted faster than the insulated ice underlying the channel-fill deposits (Fig. 9). Elongate plugs of ice that persisted beneath the channel sediments apparently acted as "sediment separators" (Fig. 9c). Ablational and superglacial debris slid down both sides of these ice cores and into cross joints to form either a

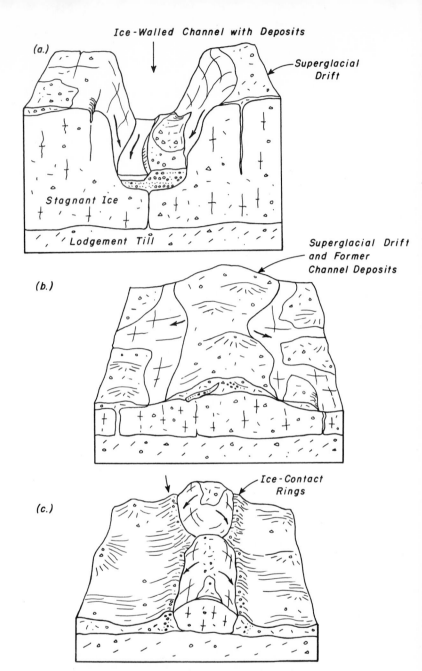

Figure 9. Probable mechanics of formation of chainlike ice-contact ridges and rings. (a) Glaciofluvial and superglacial drift accumulate in ice-walled valleys as downwasting continues. (b) The ice-walled melt-water channel is abandoned and the drift infilling retards downwasting which proceeds faster on the valley walls than beneath the valley floor. (c) The protected core of ice promotes the final inversion of topography in which channel sediments are "split" into rings and ridges. Where inversion exceeds 50 or more feet, landforms are randomly distributed and aligned rings and ridges are rarely formed.

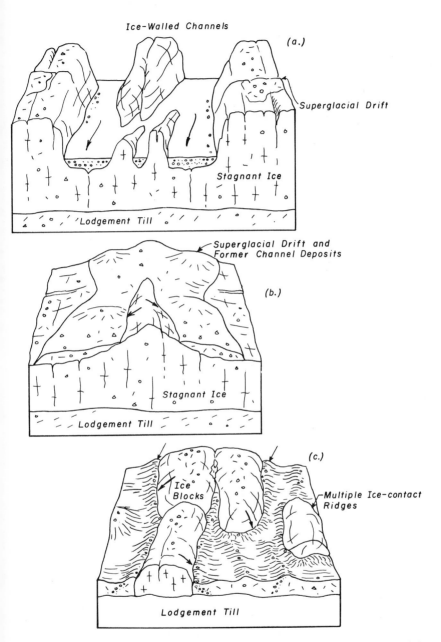

Figure 10. Probable mechanics of formation of multiple ice-contact ridges. (a) Multiple ice-contact rings and ridges produced in a similar manner as shown in Figure 9. Either multiple ice valleys localized melt water prior to a single inversion or (b) the inversion process was repeated more than once to produce (c) multiple rings and ridges.

chain-linked pattern of interconnecting ice-contact rings or multiple ice-contact ridges. Some of the rings are aligned in distinct meander patterns probably inherited from meandering channels developed on the relatively flat glacial surface (Fig. 8).

The meandering pattern of chainlike ice-contact rings and ridges, the hanging-valley relation, contorted bedding or lack of bedding in sand and gravel comprising many rings and ridges, and their frequent occurrence where dead-ice deposits and landforms are well-developed support an ablational dead-ice origin for these features.

Chainlike, ice-contact rings and branching ice-contact ridges often merge into a single ridge that is composed largely of sand and gravel. Where the depositional surface is preserved, flat-crested valley-fill sediments were deposited directly on lodgement till and were not inverted or contorted as the adjacent ice melted except along their margins. Undisturbed ice-contact ridges of this type are usually single-crested and occur in areas that are 5 to 30 feet higher in elevation than adjacent areas where the same ridges are multiple or chainlike. Floors of these valleys were eroded to the till surface beneath the base of the glacier along these valley segments and glaciofluvial deposits were not disturbed or inverted by cores of ice during the final wasting of the glacier. These channel segments were ice-walled valleys. This is not true of the sediments that make up the ice-contact rings and multiple ridges along other segments of the same valleys.

Single ice-contact ridges often merge into two or more ridges that branch off in various directions. On Plate 7, all the ridges trend essentially in the same direction and may indicate systematic joint control parallel to the direction of ice advance as indicated by their right-angle relation to end moraines. Clayton (1967, editorial comments) offered an alternative suggestion that melt water may have flowed down an even ice slope with no joint control to produce a similar relation.

Multiple ice-contact ridges, or ridges with more than one crest, undoubtedly were formed by variations of the process described above. Either interconnected and branched ice-walled valleys were present as indicated by the branched and interconnected ridges observed, or crevasse fill deposits split into one or more discontinuous, parallel ice-contact ridges during the inversion process. It is possible that channel deposits were inverted more than once by an ice core in the cases where sediments were located higher up in the ice or they were split into several ridges by underlying ice blocks that were initially more highly jointed (Fig. 10). The former explanation is favored for some multiple ice-contact ridges because the ridges often merge into hummocks, dimpled knobs, crenulated knobs, or rimmed kettles where the ridge trends are projected into local depressions. This suggests that the linear trend of ice-contact ridges was destroyed where excessive relief was involved in the inversion process.

This relation accounts for the lack of an ice-contact ridge or chainlike rings to the east of point A (Pl. 8). Opposite this hanging outwash plain where its elevation is highest, rimmed kettles composed largely of glaciofluvial sediments were produced rather than ice-contact ridges. Also, ice-contact ridges frequently tend to give rise to knob-and-kettle and rimmed-kettle topography as they are traced where previously thicker ice stood. This indicates that the elevation of debris above the base of the stagnant glacier controlled whether single or multiple ice-contact ridges and chainlike rings were produced or whether less regular hummocky dead-ice landforms developed.

Final form of ice-contact ridges must have depended upon the relief involved in the inversion process, the amount of debris present as fill in the ice-floored valleys, degree to which melt water continued to rework the valley fill sediments prior to inversion, and the amount and distribution of debris within the stagnant ice.

Some ice-contact ridges undoubtedly are true eskers that were formed in subglacial tunnels. This origin, however, does not satisfactorily explain the chainlike appearance of ice-contact rings and ridges, the tight meander pattern of some rings and ridges, their relation to hanging valleys, or their association with moraine plateaus, partly filled melt-water channels, and open melt-water channels.

Lemke (1960, p. 80–81) described similar landforms in the Souris River area of North Dakota that are referred to as one variety of ice-contact rings and ridges in this report. These were classified as "eskers" and "composite eskers" consisting of several sinuous ridges partly coalesced. These were mapped on the southeastward extension of the Missouri Coteau upland and have size, shape, extension, and composition similar to genetically related features in the Willow Bunch Lake area. In Ward County, North Dakota, "coalesced eskers" are especially well-developed, according to Lemke. They arise as a series of subparallel ridges that unite into a rectilinear network of ridges in a distance of less than one-half mile. Nearby are nearly parallel sinuous ridges with indistinct, canallike intervening troughs. R. B. Colton (oral commun. to Lemke, 1960, p. 81) speculated on an origin similar to that presented in this report for ice-contact ridges formed in ice-walled valleys.

He speculated that both ridges were deposited in a single superglacial channel on ice only a few feet thick. Ice on both sides of the channel melted before the protected ice melted underneath the deposit in the floor of the channel. Part of the deposit then slid to the ground on one side of the former channel and the remainder slid down the other side, thus forming individual ridges. The ice that had been under the channel then melted away, leaving the intervening trough.

Colton's explanation and the one presented here are in agreement with the exception that the ice thickness beneath the original valley fill is known to have exceeded 30 feet or more in some cases in the Willow Bunch Lake area. Multiple inversions from higher positions within the dead ice and the mechanism for the formation of chainlike ridges are believed to be new.

Similar ice-contact ridges and chainlike rings have been referred to as "double disintegration ridges" by Clayton (1967, Fig. A-5). His explanation for their origin is similar to that of Gravenor and Kupsch (1959, p. 52–54) and to mine. However, little evidence can be cited for an ice-press mechanism to explain the origin of these features in the Willow Bunch Lake area, particularly where these features are composed almost entirely of sand and gravel and where they transcend ground moraine composed largely of till. This would require the ice-pressed debris to be reworked by melt water at the base of the glacier as it was pressed into a distinct ridge.

Ice-floored valleys that appeared to be little modified by melt water and contained little if any glaciofluvial sediments produced similar low-relief ice-contact ridges and chainlike rings that are composed largely of till. The only distinction between these landforms and the adjacent hummocky ground moraine is the fact that they display a linear trend.

The relief on ice-contact chainlike rings and ridges increases as does the relief of adjacent landforms and the thickness of drift comprising these stagnant-ice landforms. Thickness of dead-ice deposits and relief on ice-contact rings and ridges both increase as ice-thrust moraines and end moraines are approached, indicating that ice-contact ridges and rings are best developed where englacial and superglacial debris was abundant within the glacier. This observation favors a letdown origin for ice-contact ridges rather than an ice-press mechanism. Where glaciers stagnated on a regional basis, one would expect a more nearly uniform distribution of ice-contact ridges and rings if they were produced by an ice-press mechanism rather than a transitional and restricted relation for their distribution as is observed in the Willow Bunch Lake area. Rather unique and localized subglacial conditions would have to be called upon to explain their distribution if they had largely an ice-press origin. Similarly, these landforms should be well-developed in regions containing hummocky dead-ice topography as well as in adjacent gently undulating ground-moraine areas where it can be shown that glacial ice stagnated over the entire region. This is not the case in the Willow Bunch Lake area where massive stagnation can be demonstrated as in Coronach lowland in the southwestern corner of my area (Pl. 1).

Ablational deposits along the margins of moraine-lake plateaus 100 to 150 feet above the general level of the adjacent hummocky dead-ice moraine, ablational debris capping ice-thrust bedrock features up to 500 feet higher in elevation than adjacent ground moraine on the Missouri Coteau upland in the

Willow Bunch Lake area, hanging outwash deposits, and similar observations indicate that an abundance of superglacial and englacial debris was available within the ice to give rise to these landforms by an ablational process during downwasting of the ice surface.

Ice-contact ridges might be classified simply as crevasse fillings by some workers. However, even though some ice-contact ridges and some ice-contact rings originated in crevasse environments, others have a distinctly different pattern and appearance from crevasse fillings of the type described by Ellwood (1961, 50–73), Christiansen (1956, p. 15, 1959, p. 16), and others. The unique appearance of some ice-contact rings and ice-contact ridges excluding crevasse fillings and eskers seems to be related to the effects of glaciofluvial action and to the effects of an inversion process that was involved in their late stage of formation. The crevasse fillings illustrated by Ellwood (1961), on the other hand, appear to have been deposited directly by gravity at the base of open crevasses that extended to the base of the glacier. These crevasse fillings do not appear to have been involved in an inversion process except as they were let down along their margins by melting ice or let down as a single coherent mass.

Ice-Contact Ridges Marginal to Outwash Plains

Ice-contact ridges are associated with the margin of the re-entrant outwash plain north of Horizon (9, 17, and 20–7–24) (Pls. 1, 8) and along the southern margin of the outwash plain at Ormiston (8–24) (Pl. 1). These ridges have 5 to 20 feet of relief and are composed principally of sand and gravel. Examples are shown on Plate 8 and Figure 11. Some of the ridges delineate intralobate re-entrants in what was the margin of the ice. Where two or more ridges were parallel to the ice margin they may mark successive positions of the ice front as it receded or they may have formed as multiple crevasse fillings parallel to the ice margin. Where these ridges have highly irregular crests they may be confused with kame moraines. Other ridges, composed of sand and gravel, were deposited in crevasses and ice-walled feeder valleys at the apex of the outwash plain at right angles to the glacier's margin.

The main body of outwash at Ormiston was deposited on thin, stagnant ice. The hummocky character of the outwash plain was acquired as the ice melted from beneath the sand and gravel. Knobs and kettles, ice-contact rings and ridges, dimpled knobs, and most of the other landforms that occur in hummocky and transitional dead-ice moraines were formed by a combined inversion and ablation process, but, in this case, landforms are composed principally of ablational sand and gravel rather than till. The superglacial and ablational till that was originally present in and on the surface of the waning ice was reworked or swept away by melt water.

ICE-CONTACT RIDGES ASSOCIATED WITH RE-ENTRANT
OUTWASH PLAIN NORTH OF HORIZON, SASKATCHEWAN

The "washed surface" outlines the hummocky outwash plain (HOP). It is bordered by ice-contact ridges (ICR) to the east and by multiple ice-contact ridges marking feeder melt-water channels. The outwash plain is surrounded by hummocky ground moraine (HGM) and a partly buried melt-water channel (BC) to the west. Point A marks the apex of a feeder channel. Dimpled knobs and rimmed kettles to the east of point A indicate that the inversion process has completely destroyed the trend of the original ice-walled valley. Sand, gravel, and an abundance of boulders mixed with till in these knobs reveal the former channel position. The eastern face of the outwash plain is hanging 10 to nearly 75 feet with respect to hummocky ground moraine (15, 16, 17, 20, 21, 22–7–24) and indicates minimum relief in the inversion process.

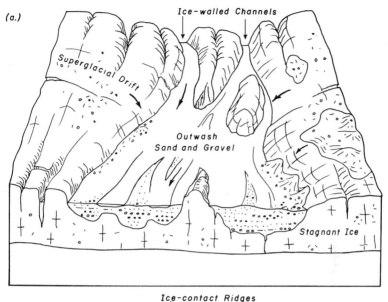

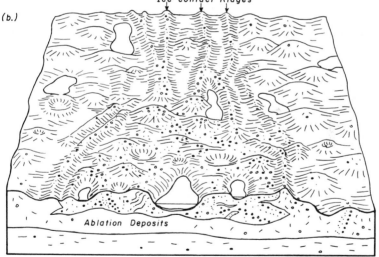

Figure 11. Probable environment of formation of ice-contact ridges associated with hummocky outwash plain near Ormiston and Horizon, Saskatchewan. (a) Superglacial outwash is deposited where re-entrants in the ice front localize melt water. (b) Ice-contact ridges, multiple ridges, or rings form along the feeder channels and crevasses, depending upon the floor elevation of the ice-walled valleys. Ice-contact ridges form marginal to the ice surface on either side of the re-entrant. These may be one or more in number, indicating parallel joint control and/or may mark successive positions of the ice front.

Plate 8 shows the apex of the outwash plain near Horizon. The hummocky character of the outwash plain is not as well-developed in this region as it is at Ormiston. Ice-contact ridges are best seen in stereo; hence they have been outlined so they might be seen. The tonal contrast between the till in the hummocky ground moraine and the sand and gravel of the outwash plain and the washed appearance of the constructional surface are well shown.

Ice-Contact Ridges Marginal to Hummocky Ground Moraine

Ice-contact ridges, several hundred feet to more than 4 miles in length, outline some areas of hummocky ground moraine (14, 21, and 22–11–23, 19, 29, and 30–9–22, 14, 21; Pl. 9 and Fig. 12). These features have been called "dump ridges" (Pettyjohn, 1967). They are associated with hummocky ground moraines that contain moraine-lake plateaus and other ice-contact rings and ridges. They may be composed of till with scattered pebble orientations characteristic of ablational deposits and chaotic assemblages of contorted lake clay, sand, gravel, and till. These features combined indicate that this form of ice-contact ridge is also of an ablational stagnant-ice origin.

The ridges may separate hummocky ground moraine from gently undulating ground moraine, transitional ground moraine, lake basins, and areas of drift swept by melt water. The relief, composition, and surface expression of these ridges are similar to other ice-contact ridges described earlier but formed under their own unique circumstances.

"Dump" ice-contact ridges outline areas where buried dead-ice masses persisted long after adjacent ground moraine had been deposited, while immediately adjacent areas were being swept by melt water or winnowed by wave action where lakes were partly dammed by stagnant ice. Each of these associations have been observed in the Willow Bunch Lake area.

Characteristically, areas of gently undulating and transitional ground moraine appear to have been deglaciated before areas of hummocky ground moraine and almost always contain thinner drift than adjacent hummocky ground moraine. This, combined with the greater relief of hummocky ground moraine, indicates that greater amounts of ablation drift were present higher in the ice and served as insulation to underlying masses of stagnant ice. The stagnant ice containing more debris persisted longer than the relatively cleaner ice in adjacent areas. The ice-contact ridges in question marked the transitional boundary between the cleaner and dirtier ice that was exposed during late-stage melting (Fig. 12b). A mechanism by which ablational and superglacial debris sloughed off buried stagnant ice onto exposed ground moraine and abandoned lake sediments adequately accounts for all features observed.

Ridges of this type border hummocky dead-ice moraine on the Missouri Coteau upland in some areas and separated dead-ice moraine from the meltwater–scoured bedrock and ground-moraine surfaces of the northeastern face

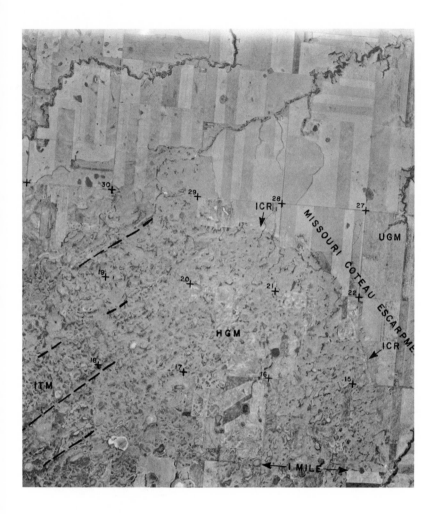

ICE-CONTACT RIDGE MARGINAL TO HUMMOCKY GROUND
MORAINE OF STAGNANT-ICE ORIGIN

Ice-thrust moraine of low relief (ITM) and hummocky ground moraine are abruptly truncated along the Missouri Coteau escarpment. Ice-contact ridges (ICR) mark the line of truncation. Gently undulating ground moraine (UGM) corresponds to a region covered by thin drift 5 to 30 feet compared with 50 to 300 feet of drift in the hummocky ground-moraine region (11–23).

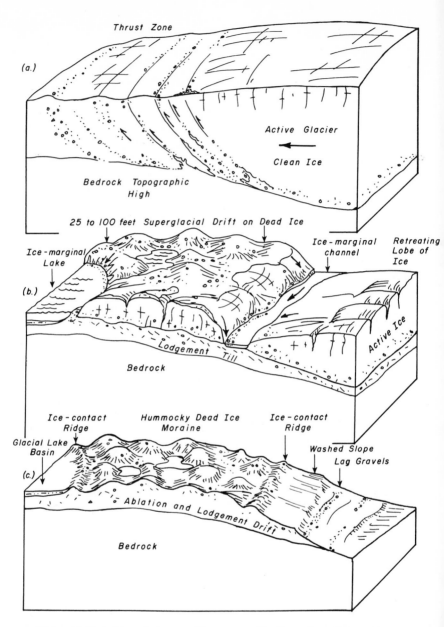

Figure 12. Probable environment of ice-contact ridge formation adjacent to ice-dammed lakes, adjacent to melt-water raceways developed against stagnant ice, and in the transitional zone between hummocky ground moraine of stagnant-ice origin and gently undulating ground moraine. (a) Thrusting of live ice against stagnant ice enriches the glacier with debris. (b) During melting superglacial drift accumulates to retard downwasting. Ice-marginal channels and lakes, superglacial lakes, and moraine-lake–plateau basins develop. (c) Ice-contact ridges develop in contact with the wasting stagnant ice that may linger 1000 to 2000 years after ice retreats from adjacent lowlands. The lowlands may be 100 to 1000 feet lower in elevation, contain cleaner ice, and be deglaciated sooner than the uplands capped by dirty ice.

of the Missouri Coteau escarpment (Pl. 9). The main lobe of ice that was melting back from the Missouri Coteau upland after depositing the Maxwellton and Ardill End Moraines (Pl. 1) (Parizek, 1964, p. 30) appears to have retreated down the Coteau escarpment while stagnant ice persisted beneath thick superglacial drift on the Missouri Coteau upland. The main lobe separated from the stagnant ice by differential melting while melt water scoured first the dead ice on the upland and later exposed ground moraine as ice continued to retreat (Figs. 12b, 12c). The resulting ice-contact ridges formed after melt-water scour ceased to erode the dead-ice mass or they would have been swept away. Dead ice had to occupy topographically low areas in the face of the Coteau escarpment at this time or else these lows would have been modified by melt water or filled with or contain lake deposits. The fact that areas 200 to 500 feet lower in elevation to the northeast of the Coteau escarpment were being deglaciated while dead ice still survived beneath thick ablation deposits in uplands on the Coteau is not unreasonable. C^{14} dates on sediments from the Coteau upland indicate that dead ice lingered at least 2000 years under ablation deposits in the Galilee Junction–Crane Valley area (Pl. 1) while tracts of land had already been deglaciated at least 12 miles to the northeast in the direction of regional ice retreat. Peat beds were accumulating in kettle lakes in this area abandoned nearly 2000 years earlier by the retreating glacier (Parizek, 1964, p. 41). A similar slow rate of melting of buried dead ice (2000 to 3000 years) was reported by Clayton (1967, p. 36–38) to the south on the Coteau in Ward County, North Dakota.

Locally, where blocks of stagnant ice afforded closure for glacial lakes, ice-contact ridges mark the temporary boundary of the lake basin and stagnant ice. Locally, hummocky ground moraine is lower in elevation than the associated lake deposits, indicating that the ice-contact ridges were contemporaneous to the lake sediments and that the lake drained before residual ice blocks finally wasted away. On Plate 5, ice-contact ridges mark the northern slope of the Maxwellton End Moraine. A proglacial lake formed between the northward retreating lobe of ice and the end moraine with its enclosed masses of stagnant ice. Ice-contact ridges mark the southern shoreline of this lake and were formed at the margin of the lake or after the lake drained.

The origin for ice-contact ridges formed in this particular environment was proposed by Parizek (1961) and has since been supported by Pettyjohn (1967). Pettyjohn refers to these same features as "dump ridges." He indicates that dump ridges are one of several types of "disintegration ridges" that are relatively common on the Missouri Coteau in Ward County, North Dakota. They generally consist of boulders and coarse gravel that have an eskerlike form and were deposited at the edges of stagnant ice fields and thus mark positions of the ice margins. He concludes that the ridges were formed by the dumping of outwash at the margin of stagnant ice. The coarse material formed a ridge whereas finer material was carried downslope, forming adjacent out-wash deposits. A typical "dump ridge" in Ward County is 20 to 45 feet wide at the base, 5 to 20 feet in height, and as much as 1 mile in length. Where two dump ridges are parallel and closely spaced he indicates that they reflect successive retreats of the margin of the stagnant ice field.

My interpretation agrees entirely with that of Pettyjohn. In the Willow Bunch Lake area, however, ice-contact ridges formed in this setting are also composed of till, laminated silts and clays, sand and gravel, and mixtures of any of the above, depending upon the local setting.

FINAL STATEMENT

Several varieties of ice-contact rings and ridges have been illustrated and their environment and mechanics of formation proposed. In all cases the rings and ridges bordered blocks of stagnant ice as from 10 to more than 500 feet of stagnant ice wasted in place (Parizek, 1961, p. 129–134).

Ice-contact rings and ridges are best displayed in size, form, diversity, and frequency of occurrence where englacial debris was abundant, high above the base of the glacier, and where thick accumulations of superglacial debris resulted from stagnation. In the Willow Bunch Lake area, these conditions were met where live ice thrust upon and overrode dead ice near the glacier's margin during successive readvances and where thrusting at the snout of the glacier accompanied ice movement over selected topographically high areas. Significant volumes of englacial debris were lifted by this process at least 500 feet above the base of the glacier that deposited the Maxwellton and Ardill End Moraines and that formed the ice-thrust Dirt and Cactus Hills (Parizek, 1961) (Pl. 1). Clayton (1967, p. 38–41) expressed an identical opinion to account for the origin of englacial debris in ice that capped the Missouri Coteau to the south in North Dakota.

Ice-contact rings and ridges are poorly developed or absent where abla-tion drift was thin or where drift was concentrated within the lower 10 to 20 feet of the stagnant ice. This observation further suggests that an ablational process dominated in the formation of ice-contact rings and ridges in question.

The distribution, nature, and origin of some ice-contact rings and ridges also depend upon the topographic configuration of the subglacial surface in the case of rim-ringed moraine-lake plateaus, ridges marginal to partly buried valleys, and ridges marginal to masses of hummocky dead-ice moraine. Dis-tribution of others depended upon the presence of stream trenches, subglacial channels, and ice-floored crevices. Glaciofluvial, glaciolacustrine, and till de-posits gave rise to ice-contact rings and ridges in each of these environments.

It is reasonable to assume that an ice-press mechanism may have operated at the margins of stagnant ice blocks and that this may have contributed lodgement till to the base of some rings and ridges. Limited fabric studies do not suggest that lodgement till contributed to these features, at least near the upper 5 to 25 feet of their crests. It is concluded that they were produced largely by a letdown process. Additional fabric studies, however, are warranted near the base of these landforms.

Only recently the significance and specific mode of origin for several of these features have been recognized more widely by geologists. This knowledge should prove useful in interpreting the origin of adjacent landforms and deposits whose origins are still in doubt and should aid in the interpretation of the nature and history of ice-retreat in other areas. The relation between ice-contact rings and ridges and stagnant ice cannot be disputed.

REFERENCES CITED

Bik, M. J. J., 1967, On the periglacial origin of prairie mounds: North Dakota Geol. Survey Misc. Series 30, p. 83–94.

Bluemle, J. P., 1965, Geology and ground water resources of Eddy and Foster Counties, North Dakota, part 1—geology: North Dakota Geol. Survey Bull. 44, 66 p.

Bluemle, J. P., Faigle, G. A., Kresl, R. J., and Reid, J. R., 1967, Geology and ground water resources of Wells County, part 1—geology: North Dakota Geol. Survey Bull. 51, 39 p.

Christiansen, E. A., 1956, Glacial geology of the Moose Mountain area, Saskatchewan: Saskatchewan Dept. Mineral Resources Rept. 21, 35 p.

—— 1959, Glacial geology of the Swift Current area, Saskatchewan: Saskatchewan Dept. Mineral Resources Rept. 32, 62 p.

—— 1960, Geology and groundwater resources of the Qu'Appelle area, Saskatchewan: Saskatchewan Research Council Geol. Div. Rept. 1, 53 p.

Clayton, L., 1962, Glacial geology of Logan and McIntosh Counties, North Dakota: North Dakota Geol. Survey Bull. 37, 84 p.

—— 1967, Stagnant-glacial features of the Missouri Coteau in North Dakota: North Dakota Geol. Survey Misc. Series 30, p. 25–46.

Clayton, L., and Cherry, J. A., 1967, Pleistocene superglacial and ice-walled lakes of west-central North America: North Dakota Geol. Survey Misc. Series 30, p. 47–52.

Clayton, L., and Freers, T. F., 1967, Glacial geology of the Missouri Coteau and adjacent areas: North Dakota Geol. Surv. Misc. Series 30, p. 1–24.

Deane, R. E., 1950, Pleistocene geology of Lake Simcoe district, Ontario: Canada Geol. Survey Mem. 256, 108 p.

Ellwood, R. B., 1961, Surficial geology of the Vermilian area, Alberta: Urbana, Illinois Univ. Ph.D. dissert.

Flint, R. F., 1928, Eskers and crevasse fillings: Am. Jour. Sci., v. 15, p. 410–416.

Gravenor, C. P., 1955, The origin and significance of prairie mounds: Am. Jour. Sci., v. 253, p. 475–481.

Gravenor, C. P., and Bayrock, L. A., 1956, Stream-trench systems in east-central Alberta: Research Council of Alberta Prelim. Rept. 56–4.

Gravenor, C. P., and Kupsch, W. O., 1959, Ice-disintegration features in Western Canada: Jour. Geology v. 67, no. 1, p. 48–64.

Hansen, D. E., 1967, Geology and ground water resources Divide County, North Dakota, part 1—geology: North Dakota Geol. Survey Bull. 45, 90 p.

Harrison, W., 1963, Geology of Marion County, Indiana: Indiana Geol. Survey Bull. 28, 78 p.

Henderson, E. P., 1952, Pleistocene geology of the Watino quadrangle, Alberta: Bloomington, Indiana Univ. Ph.D. dissert.

Hoppe, G., 1952, Hummocky moraine regions, with special reference to the interior of Norbotten: Geog. Annaler, v. 34, p. 1–72.

—— 1957, Problems of glacial morphology and the ice age: Geog. Annaler, v. 39, no. 1, p. 1–17.

—— 1959, Glacial morphology and inland ice recession in northern Sweden: Geog. Annaler v. 41, no. 4, p. 193–212.

—— 1963, Subglacial sedimentation, with examples from northern Sweden: Geog. Annaler, v. 45, no. 1, p. 41–49.

Kume, J., and Hansen, D. E., 1965, Geology and ground water resources of Burleight County, North Dakota, part 1—geology: North Dakota Geol. Survey Bull. 42, 111 p.

Lemke, R. W., 1960, Geology of the Souris River area North Dakota: U.S. Geol. Survey Prof. Paper 325, 138 p.

Parizek, R. R., 1961, Glacial geology of the Willow Bunch Lake area: Urbana, Ph.D. dissert., Illinois Univ.

—— 1964, Geology of the Willow Bunch Lake area (72-H) Saskatchewan: Saskatchewan Research Council Geology Div. Rept. no. 4, 48 p.

Pettyjohn, W. A., 1967, Dump ridges and collapsed sub-ice channels in Ward County, North Dakota: North Dakota Geol. Survey Misc. Series 30, p. 115–116.

Royse, C. F., and Callender, E., 1967, A preliminary report on some ice-walled lake deposits (Pleistocene), Mountrail County, North Dakota: North Dakota Geol. Survey Misc. Series 30, p. 53–62.

Sproule, J. C., 1939, The Pleistocene geology of the Cree Lake Region, Saskatchewan: Royal Soc. Canada Trans., 3d ser., v. 33, sec. 4, p. 101–109.

Stalker, A. Mac S., 1960, Ice-pressed drift forms and associated deposits in Alberta: Canada Geol. Survey Bull. 57, 38 p.

Tanner, V., 1914, Studier öfver Kvartärsystemet i Fennoskandias nordliga delar. III. Om landisens rörelser och afsmältning i Finska Lappland och angränsande trakter: Finlande Comm. Géol. Bull. 38.

Winters, H. A., 1963, Geology and ground water resources of Stutsman County, North Dakota, part 1—geology: North Dakota Geol. Survey Bull. 41, 84 p.

Witkind, I. J., 1959, Geology of the Smoke Creek-Medicine Lake-Grenora area, Montana and North Dakota: U.S. Geol. Survey Bull. 1073, 80 p.

GEOLOGICAL SOCIETY OF AMERICA, INC.
SPECIAL PAPER 123

Pleistocene Niche Glaciers and Proto-Cirques, Cataract Creek Valley, Tobacco Root Mountains, Montana

ALAN M. JACOBS
Illinois State Geological Survey, Urbana, Illinois

ABSTRACT

In the valley of Cataract Creek in the Tobacco Root Mountains, Montana, two moraines were deposited by Pleistocene niche or slab glaciers that emanated from poorly developed cirques (proto-cirques) along the sides of the valley at elevations of less than 8000 feet (2438 m) rather than by valley glaciers that emanated from well-formed cirques at elevations of 9000 feet (2743 m). The niche or slab glaciers seem analogous to those occurring in presently glacierized proto-cirques in Spitsbergen, Norway, and Scotland. Proto-cirques were formed in hollows at elevations of less than 8000 feet (2438 m) in this valley because the hollows were hemmed in and orographically influenced by valley glaciers on the south and east and by the main divide of the mountains on the north and west. The two valley glaciers generated a locally cold microclimate in the hollows, and their nearness to the main divide resulted in the hollows' being well-situated for the accumulation of snow drifting across the divide. The fortuitous position of these hollows during the late Pleistocene resulted in incipient glacier formation at elevations lower than would be expected.

CONTENTS

103

INTRODUCTION

Two surficial deposits, called the Mountain Meadow and Lone Wolf deposits, are located high on the northern slope of the valley of Cataract Creek, Tobacco Root Mountains, Montana (SE¼, Sec. 21, T.2S., R.3W., Harrison 15' quadrangle; Figs. 1, 2). They have morainelike loops and cirquelike source areas that have elevations anomalously low compared to those of regional cirques.

ACKNOWLEDGMENTS

This paper is adapted from a doctoral dissertation submitted to Indiana University. I wish to thank William D. Thornbury, Herbert E. Wright, Jr., John de la Montagne, John B. Patton, and Judson Mead for stimulating discussions and encouragement. Logistic help and the field assistance of Karl F. Frey were extended through the facilities of the Indiana University Geologic Field Station in Cardwell, Montana. Funds were furnished under a grant for recipients of National Aeronautics and Space Administration traineeships. Herbert E. Wright, Jr., Jerry A. Lineback, and Gerald M. Richmond critically reviewed the final manuscript.

MOUNTAIN MEADOW AND LONE WOLF DEPOSITS

The Mountain Meadow deposit (Pls. 1, 2) has three major and several minor arcuate ridges separated by canoe-shaped depressions. The deposit is 3000 feet (914 m) long and 500 feet (152 m) wide, and its front is 50 feet (15 m) high. Ridges stand 10 feet (3 m) above adjacent depressions. The deposit is composed of unsorted angular cobbles and boulders in a matrix of finer materials to a depth of at least 12 feet (4 m). The long axes of cobbles have a pronounced orientation N. 60° to 90° W., perpendicular to the inferred direc-

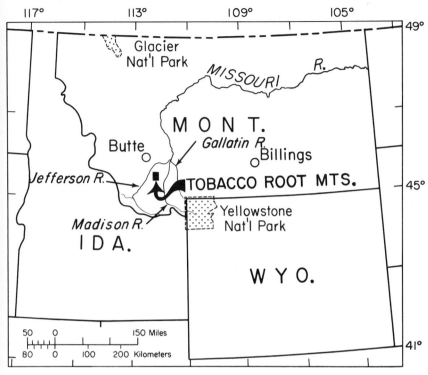

Figure 1. Index map showing location of Tobacco Root Mountains.

tion of movement of the mass. The deposit overlies igneous and metamorphic bedrock (Fig. 2). Pebble counts within the terminal ridges yielded a mean of 58.0 ± 4.7 percent[1] igneous rock.

The Mountain Meadow deposit can be traced headward into a tree-covered hollow (NE¼, Sec. 29, and NW¼, Sec. 28; Fig. 2 and Pl. 1) from which the deposit was presumably derived. The outer rim of the hollow lies about 6000 feet (1829 m) away from the terminal ridge of the deposit. The northeast-facing hollow is underlain only by metamorphic bedrock and is covered with a thin colluvial soil.

Because (1) the hollow is underlain only by metamorphic bedrock, (2) the deposit overlies igneous and metamorphic bedrock, and (3) the terminal ridges of the deposit contain a mixture of igneous and metamorphic rock, erosion and incorporation of bedrock from beneath the terminal area of the deposit is strongly suggested.

---

[1] Level of confidence on all pebble counts is 95 percent.

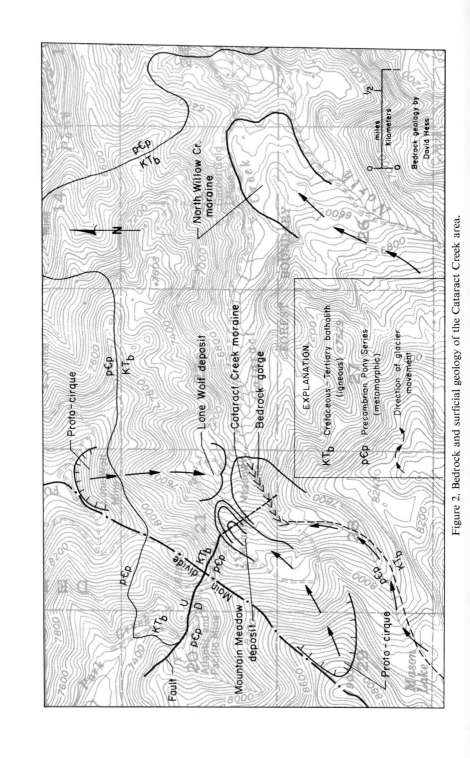

Figure 2. Bedrock and surficial geology of the Cataract Creek area.

PANORAMA OF THE MOUNTAIN MEADOW DEPOSIT AND ITS SOURCE AREA

TERMINAL LOOPS OF THE MOUNTAIN MEADOW DEPOSIT

CATARACT CREEK MORAINE

MOUNTAIN MEADOW DEPOSIT

JACOBS, PLATE 2
Geological Society of America Special Paper 123

The Lone Wolf deposit lies along the valley side downstream from the terminal Mountain Meadow ridge (NE¼, SE¼, Sec. 21; Fig. 2). The deposit is 2000 feet (610 m) long and 3000 feet (914 m) wide, and its front is 50 feet (15 m) high. It has two arcuate ridges and irregular, hummocky masses of debris. The deposit is composed of unsorted angular cobbles and boulders in a matrix of finer materials. Only igneous bedrock underlies the deposit (Fig. 2).

The concave sides of the ridges of the Lone Wolf deposit face a hollow on the northern valley wall (NE¼, Sec. 21, and SE¼, Sec. 16) from which the deposit was presumably derived. The outer rim of the hollow lies 5000 feet (1524 m) away from the terminal ridge of the deposit. The headwall of the hollow is underlain by metamorphic bedrock, and the steep, south-facing slope between the headwall and the deposit is underlain by igneous bedrock covered with a thin colluvial soil.

ADJACENT DEPOSITS

An end moraine that fills the main valley of Cataract Creek (SE¼, Sec. 21; Fig. 2) lies adjacent to the Mountain Meadow and Lone Wolf deposits. Cataract Creek flows within a bedrock gorge along the southern valley wall. Debris of the end moraine is remarkably similar to that of the Mountain Meadow and Lone Wolf deposits. Pebble counts yielded 70.5 ± 4.3 percent igneous rock. The bedrock beneath the main valley is igneous and metamorphic, although the relative amount of each is difficult to ascertain (Fig. 2).

About 2 miles downstream from the Cataract Creek end moraine, where Cataract Creek joins North Willow Creek, an end moraine derived from the valley of North Willow Creek blocks the valley of Cataract Creek (SE¼, Sec. 23; Fig. 2). This end moraine resembles the Cataract Creek end moraine in form but is much larger.

ORIGIN OF THE MOUNTAIN MEADOW DEPOSIT

The Mountain Meadow deposit might be: (1) landslide debris, (2) stabilized rock glaciers, (3) stabilized solifluction lobes, (4) protalus ramparts, (5) moraines deposited by former valley glaciers of the Cataract Creek Valley, (6) moraines deposited by outlet glaciers from an ice cap that covered the main divide of the Tobacco Root Mountains, or (7) moraines deposited by niche or slab glaciers that formed in tributary hollows of the Cataract Creek Valley.

A landslide origin is rejected because of extensive erosion beneath the terminus of the Mountain Meadow deposit. The headwall area for the Mountain Meadow deposit is a hollow underlain by only metamorphic bedrock. Only the terminus of the deposit overlies igneous bedrock. Yet igneous rock comprises 58.0 ± 4.7 percent of the debris at the terminus of the Mountain Meadow deposit. Erosion by landslides, moreover, is largely confined to the source headwall, which in this case is metamorphic and should produce debris containing little igneous rock.

The morphology of the Mountain Meadow deposit is different from typical landslide morphology, according to criteria presented by Ritchie (1958) and Varnes (1958). The arcuate shape of its ridges and the canoe shape of its depressions are not common to any type of fall or slide debris. Although landslides such as earth flows, sand flows, and silt flows can be lobate, they are composed of only fine debris, unlike the Mountain Meadow deposit.

The Mountain Meadow deposit has many features in common with rock glaciers: general lobate form, parallel V-shaped ridges and furrows, angular debris, and a steep front and sides. These features reflect the similarity between rock glaciers and true glaciers and the fact that some rock glaciers have cores of true glacier ice (Outcalt and Benedict, 1965; Potter, 1968). Insulation of glacier ice by a debris mantle can account for glacier persistence at lower than expected elevations. Ice-cored rock glaciers in the Absaroka Mountains, Wyoming (N. Potter, 1968, written commun.), now occur below the level suitable for glacier formation. Because the difference between a debris-mantled glacier and an ice-cored rock glacier is gradational and a matter of definition, the term "glacier" in this paper can include one having a thick debris mantle, which in some classifications qualifies it to be called an ice-cored rock glacier. Origin of the Mountain Meadow deposit as a debris-mantled glacier is therefore a possibility. Origin of this deposit as a rock glacier composed of rock debris with interstitial ice is rejected, however, because of extensive erosion of bedrock beneath the terminus of the deposit.

The Mountain Meadow deposit did not result from solifluction because it has a high, steep front not typical of solifluction deposits. Nor could the extensive erosion of bedrock from beneath the terminus of the deposit have been caused by solifluction. Solifluction deposits, moreover, do not have the high local relief, arcuate ridges, and canoe-shaped depressions so obviously displayed on the Mountain Meadow deposit (Pls. 1, 2).

Origin as a protalus rampart is unlikely because the Mountain Meadow deposit terminus lies 0.75 mile (1.2 km) from the foot of the headwall and contains debris eroded from bedrock beneath the terminus.

This deposit is not a moraine from a glacier in the main valley, because it can be traced out of the valley into a hollow. The headwall of the hollow is separated from the main valley by a high bedrock ridge. The absence of glacial deposits on the main divide and the high degree of dissection of upland land surfaces indicate that the deposit was not derived from an outlet glacier of an ice cap on the main divide.

The Mountain Meadow deposit is most probably till, and its ridges are most probably moraines deposited by a niche or slab glacier that formed in the tree-covered hollow in the NE¼, Sec. 29, and NW¼, Sec. 28 (Fig. 2 and

Pl. 1). Such a hollow seems analogous to present-day glacierized proto-cirques in Spitsbergen, Norway, and Scotland (Groom, 1959; Grove, 1961). A proto- or incipient cirque can support only a small niche or slab glacier. Groom (1959) noted that niche glaciers can deposit ridges of till.

ORIGIN OF THE LONE WOLF DEPOSIT

Evidence that the Lone Wolf deposit was similarly derived from a proto-cirque glacier is not as complete as that for the Mountain Meadow deposit. A similar origin is proposed, however, because of the morphology of its terminal loops and its nearness to the Mountain Meadow deposit. Because the presumed source area of this deposit is a south-facing hollow, one cannot completely discount the influence of mass movement from freeze and thaw phenomena.

CORRELATION

The glacial deposits of the Tobacco Root Mountains have been correlated by Jacobs (1967) with deposits in the Madison and Gallatin Ranges (Hall, 1960) and the Wind River Mountains (Richmond, 1965) (Table 1). The location and freshness of topography support a correlation of the Mountain Meadow and Lone Wolf deposits and the Cataract Creek and North Willow Creek end moraines with the undifferentiated Bull Lake–Pinedale glaciations. Probably the Cataract Creek and North Willow Creek end moraines are composed of Bull Lake till that was reworked and combined with till from a Pinedale glacier. The greater degree of hummocky topography at the upvalley end of both end moraines supports their composite nature.

DISCUSSION

Sixty-eight cirques in the Tobacco Root Mountains have well-formed cirque morphology: a steep headwall, a bowl-shaped depression, and a threshold. They range in elevation from 8400 to 9600 feet (2560 to 2926 m). Their mean, median, and mode elevations are 9020 feet (2749 m), 9050 feet (2758 m), and 9000 feet (2743 m), respectively. Twenty of the sixty-eight cirques have floor elevations less than 9000 feet (2743 m), and three are at 8400 feet (2560 m). Orographic snow line for the Tobacco Root Mountains, computed as the median altitude between terminal moraines and the highest points on associated cirque headwalls, is at about 8500 feet (2591 m). The elevations of the proto-cirques, at 7900 feet (2408 m) and 7500 feet (2286 m), are 1100 to 1500 feet (335 to 457 m) below the mean altitude of well-formed cirques and 600 to 1000 feet (183 to 305 m) below orographic snow line.

Proto-cirques were formed in hollows during Glaciation II (Table 1) at less than 8000 feet (2438 m) in the Cataract Creek Valley. The hollows were

TABLE 1. CORRELATION OF GLACIAL DEPOSITS

Approximate Years B.P.	Tobacco Root Mountains (Jacobs, 1967)		Madison-Gallatin Ranges (Hall, 1960)	Wind River Mountains (Richmond, 1965)
4000 and younger (Recent)	Cirque glaciers; moraines at 9000 feet (2743 m)	Glaciation III	Youngest glaciation	Neoglaciation
6500 to 45,000 (Wisconsinan)	Major valley glaciers; terminal moraines at 6500 to 7000 feet (1981 to 2134 m)		Sawmill stage	Pinedale Glaciation
	Local ice caps on two interstream divide areas on the eastern side of the range; small till sheets	Glaciation II (can be differentiated into two stades in some valleys)	Intermediate stage	Bull Lake Glaciation
	Niche and slab glaciers; proto-cirque moraines			
Greater than 45,000 (early Wisconsinan or pre-Wisconsinan)	A few lateral moraines perched on valley walls high above valley bottoms	Glaciation I	Marble Point stage	Pre-Bull Lake glaciations

initiated by headward erosion of small tributaries of Cataract Creek and were enlarged by snow-patch nivation. By the time of Glaciation II, the hollows were large enough to be protected sites for snow accumulation and glacier-ice formation. Because the hollows were hemmed in by the North Willow Creek valley glacier on the east, the Cataract Creek valley glacier on the south, and the main divide of the mountains on the north and west, a cold microclimate was generated in the hollows. As the hollows were near the main divide, they were well-situated for the accumulation of snow drifting across the divide. Local conditions of low ablation and high accumulation of snow resulted in proto-cirque development at 7900 and 7500 feet (2408 and 2286 m) in the Cataract Creek Valley.

Proto-cirques have not formed in other valleys of the Tobacco Root Range below 7500 feet (2286 m) because, besides being much lower than orographic snow line, areas at those elevations were distant from the main divide and main valley glaciers. Proto-cirques could have formed in all valleys above 7500 feet (2286 m), but at high elevations main valley glaciers would have incorporated ice and till from niche glaciers and destroyed their terminal moraine loops.

Not only was the Cataract Creek area in a good position for proto-cirque development, but it was also well-located for the preservation of morainal loops from niche glaciers. The bend in the Cataract Creek Valley (SW¼, SE¼, Sec. 21; Fig. 2) deflected the main valley glacier to the east and left intact the Mountain Meadow and Lone Wolf deposits.

The foregoing data indicate that the Mountain Meadow and Lone Wolf deposits are moraines that were formed by niche or slab glaciers in proto-cirques below the mean altitude of well-formed cirques and orographic snow line. The Cataract Creek area was hemmed in by valley glaciers and the main divide, resulting in local microclimatic conditions of low ablation and high accumulation of snow. Nivation hollows provided good sites for niche- or slab-glacier formation. As the main valley glacier was deflected to the east, it did not destroy the well-formed loops of the Mountain Meadow and Lone Wolf moraines.

REFERENCES CITED

Groom, G. E., 1959, Niche glaciers in Bünsow Land, Vestspitsbergen: Jour. Glaciology, v. 3, p. 369–376.

Grove, J. M., 1961, Some notes on slab and niche glaciers, and the characteristics of proto-cirque hollows: Internat. Assoc. Sci. Hydrology Pub. 54, p. 281–287.

Hall, W. B., 1960, Multiple glaciation in the Madison and Gallatin Ranges, southwestern Montana: Billings Geol. Soc., Eleventh Annual Field Conf., p. 191–199.

Jacobs, A. M., 1967, Pleistocene proto-cirque hollows in the Cataract Creek Valley, Tobacco Root Mountains, Montana: Bloomington, Indiana Univ., Ph.D. dissert., 95 p.

Outcalt, S. I., and Benedict, J. B., 1965, Photo-interpretation of two types of rock glacier in the Colorado Front Range, U.S.A.: Jour. Glaciology, v. 5, p. 849–856.

Potter, N., Jr., 1968, Galena Creek rock glacier, Northern Absaroka Mountains, Wyoming (Abstract): p. 438 in Abstracts for 1967, Geol. Soc. America Spec. Paper 115.

Richmond, G. M., 1965, Glaciation of the Rocky Mountains, p. 217–230 in Wright, H. E., Jr., and Frey, D. G., Editors, The Quaternary of the United States: Princeton, New Jersey, Princeton Univ. Press, 922 p.

Ritchie, A. M., 1958, Recognition and identification of landslides, Chap. 4 *of* Eckel, E. B., *Editor*, Landslides and engineering practice: Natl. Acad. Sci.–Natl. Research Council Highway Research Board Spec. Rept. 29, p. 48–68.

Varnes, D. J., 1958, Landslide types and processes, Chap. 3 *of* Eckel, E. B., *Editor*, Landslides and engineering practice: Natl. Acad. Sci.–Natl. Research Council Highway Research Board Spec. Rept. 29, p. 20–47.

GEOLOGICAL SOCIETY OF AMERICA, INC.
SPECIAL PAPER 123

The Saidmarreh Landslide, Iran

R. A. WATSON
Washington University, St. Louis, Missouri

H. E. WRIGHT, JR.
University of Minnesota, Minneapolis, Minnesota

ABSTRACT

More than 10,000 years ago a slab of Tertiary limestone 15 km long slid off the northern flank of Kabir Kuh in the Zagros Mountains of southwestern Iran, producing a landslide that crossed two valleys and an intervening ridge and extended 20 km from its source. The slide must have been triggered by an earthquake, but the limestone slab probably had previously been undercut by the Saidmarreh River. Evidence is not adequate to postulate lubrication by a layer of compressed air; instead, pulverized marl and a sliding surface of gypsum bedrock may account for the distant travel.

The surface of the slide is crisscrossed by ridges and troughs and by large grabens, mostly related to collapse over subsurface cavities produced by solution of gypsum bedrock after the slide occurred, both during outseepage of dammed lakes and, subsequently, after the lakes were drained by erosion of the outlets. The surface features resemble those formed on drift-mantled stagnant glaciers or by solution on limestone, that is, karst.

CONTENTS

INTRODUCTION

The largest landslide known on earth is the Saidmarreh on the northern flank of Kabir Kuh, the southernmost major ridge of the Pusht-i-Kuh region of Luristan in the Zagros Mountains, southwestern Iran (Fig. 1). The Kabir Kuh, with average elevation of about 2000 m above sea level, is a northwest-trending anticlinal ridge capped by massive Middle Cretaceous limestone. The northern flank features a high hogback of Asmari Limestone (Oligocene–Lower Miocene) 300 m thick. This massive unit dips about 20° and rests on thin-bedded Eocene limestone and marl. A segment of the hogback 15 km long, 5 km wide, and at least 300 m thick slid off the mountain into the valley of the Karkheh River (Pls. 1, 2). Part of the mass of broken debris had sufficient momentum to rise 600 m from the valley floor to cross the plunging nose of the anticlinal mountain to the north, coming to rest in the next valley, as much as 20 km from the point of origin (Fig. 2). The debris at the base of the mountain dammed the Saidmarreh River to form a lake 40 km long in which was deposited 125 m of sediment, and at its northernmost extension the debris dammed the Kashgan River to form a lake 9 km wide that extended an additional 11 km northward into a mountain gorge.

This great jumble of debris was attributed to Pleistocene glaciation by DeMorgan (1895, Pls. 67, 80), for it resembles a great moraine in certain respects, but Harrison and Falcon (1937, 1938) recognized its true origin and described its basic dimensions and geology. We provide new information on the fracture and depression patterns revealed by aerial photographs and field examination, a minimum date for the event, and a discussion of air-layer lubrication as a possible explanation for the slide.

ACKNOWLEDGMENTS

Field work was incidental to geomorphic and paleoecologic studies for the Iranian Prehistoric Project of the Oriental Institute, University of Chicago, directed by R. J. Braidwood and supported by a National Science Foundation grant. Maps and other facilities were loaned by the Khuzestan Development Service, Amman and Whitney Company, and Kampsax Company. Particular thanks are due to Dr. Charles Simkins of Khuzestan Development Service. R. L. Shreve kindly reviewed a draft of the manuscript.

GEOLOGIC SETTING

Symmetrical folds of the northwest-trending Zagros Mountains, southwestern Iran, expose Cretaceous limestones on anticlinal crests and preserve Miocene red beds and Pliocene conglomerates in synclines (Table 1). Kabir Kuh is the southernmost ridge in the series. It presents a steep face to the Mesopotamian piedmont for 160 km. Its crest is multiple along most of its length, for the highest stratigraphic unit, the massive Asmari Limestone (Oligocene/Miocene), is generally breached by erosion to provide two high

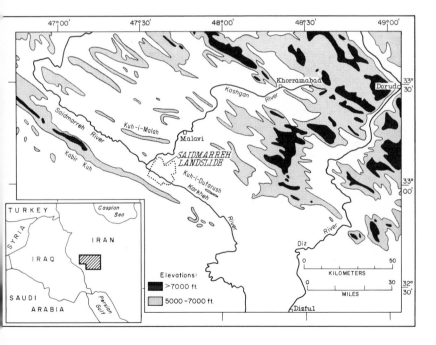

Figure 1. Map of part of Zagros Mountains, southwestern Iran. Locations of Saidmarreh landslide and of other features mentioned in the text are shown.

OBLIQUE AERIAL PHOTOGRAPH OF SAIDMARREH LANDSLIDE AND KABIR KUH

The entire scar of the Saidmarreh landslide is visible beneath the snow-capped crest of the Kabir Kuh. The central section was excavated more deeply through the thin-bedded Eocene limestones to make a paired inner scar (Pl. 1). The paired outer scar, on Asmari Limestone, is visible at the extremities of the photograph. In the near foreground is the anticlinal Kuh-i-Dufarush, whose cap of Asmari Limestone is buried by the slide

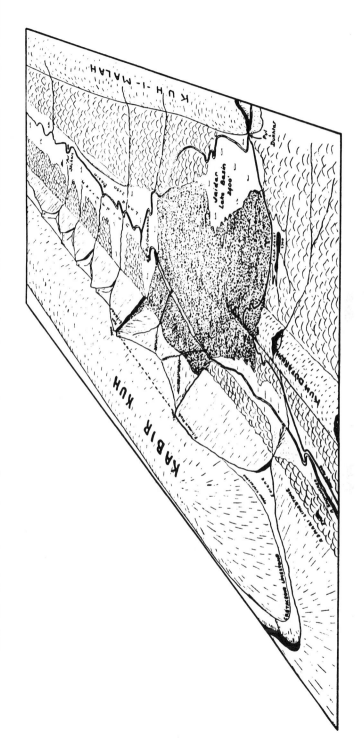

Figure 2. Block diagram of Saidmarreh landslide. *From* Harrison and Falcon (1938).

TABLE 1. BEDROCK STRATIGRAPHY OF PUSHT-I-KUH AREA, SOUTHWESTERN IRAN

	Formation	Maximum thickness (m)	Lithology
Pliocene	Upper Bakhtiari	900	Massive dark-brown conglomerate with pebbles mainly of limestone but also of chert, greenstone, and volcanics
	Unconformity		
Miocene	Lower and middle Bakhtiari Upper and middle Fars	1400	Conglomerate, sandstone, marl, clay, and gypsum
	Lower Fars	450	Gypsum, marl, clay, and limestone
	Unconformity		
Miocene/ Oligocene	Asmari Limestone	300	Massive limestone
Eocene	Middle	900	Thin-bedded limestone, marl, and conglomerate
	Lower	500	Marl, marly limestone, limestone, and shale
Cretaceous	Upper	600–900	Massive to thin-bedded limestone, shale, marl
	Lower and Middle	1500	

Slightly modified *from* Jordan (1956). *See also* Harrison and Falcon (1938).

lateral hogbacks that enclose numerous lower ridges of Eocene and Cretaceous limestones (Oberlander, 1965).

The Asmari Limestone is overlain by red beds and gypsum of the lower Fars Formation (Miocene), which floors the valley of the Saidmarreh and Karkheh Rivers to the northwest. The next major continuous anticlinal fold forms Kuh-i-Malah, a ridge of Asmari Limestone almost as high as Kabir Kuh. It has not been significantly breached along the crest. The Kashgan River cuts a sharp gorge through this and more northerly ridges to enter the Saidmarreh River just west of the landslide debris.

Kuh-i-Dufarush makes a plunging anticline between Kabir Kuh and Kuh-i-Malah. It reaches 1600 m in elevation to the east, has a smooth stripped surface on Asmari Limestone, and is transected by two symmetrical gorges cut by short tributaries to the Karkheh River superimposed from an earlier cover of the soft Fars Formation (Oberlander, 1965). This anticline plunges westward to split the Saidmarreh Valley longitudinally; the Asmari cap rock disappears westward beneath the landslide debris, but it reappears as a low ridge 4 km west of the slide area.

GROSS MORPHOLOGY OF THE LANDSLIDE DEBRIS AND ADJACENT AREAS

Landslide debris covers a roughly fan-shaped area; its proximal (southern) edge is low on the northern flank of Kabir Kuh, and its distal margin extends to the north a maximum of 16 km in two broad lobes (Pl. 1). Its surface elevation above sea level is about 1100 m at the proximal end at the base of the scar, 900 m in the Karkheh Valley, 1200 m where it crosses the nose of Kuh-i-Dufarush, and 900 m at its distal margin. It has a maximum thickness of 300 m, an area of 270 km², and a volume of almost 30 km³ (Harrison and Falcon, 1937). Its edges, where not overlapped by lake deposits, are sharp fronts generally at least 50 m high; the steepest edge is on the top of Kuh-i-Dufarush (Pl. 3, fig. 2). The northwestern front of the slide is also abrupt, as it faces a badland of dissected Fars gypsum.

The landslide mass overrode the low gypsum hills that floored the Karkheh Valley and flanked the slopes of the three anticlinal mountain ranges. The buried bedrock topography has a local relief of up to 100 m. Beneath the eastern portion of the debris and in the open valley to the east, the low gypsum hills are capped with a 10- to 30-m bed of Bakhtiari (?) dark-brown conglomerate, which rests on a 3- to 10-m bed of lower Fars limestone. The conglomerate is composed of subrounded pebbles, cobbles, and boulders up to 1 m in diameter. Most of the stones are limestone, but up to 25 percent are red, white, or purple chert, brown and yellow sandstone, some greenstone, and rare volcanic rocks. Fresh breaks in the dark-brown conglomerate along the river bank are buff colored. Chert pebbles derived from this conglomerate lie on the surfaces of a few gypsum hills that were buried beneath landslide debris and subsequently exposed by erosion. Similar pebbles occur in the sediments of the lakes dammed by the landslide, indicating that the conglomerate was extensive over the area before the slide took place.

Two major lakes now filled with sediment were impounded by landslide debris. Saidmarreh Lake extended 40 km up the Saidmarreh River to cover an area of 200 km² (Fig. 2). It was filled with as much as 125 m of sediment (Harrison and Falcon, 1937). Jaidar Lake covered an area of 90 km² north of the slide debris at the mouth of the Kashgan River. Its bottom deposits of thinly bedded, marly, clayey silt lap onto the northern margins of the slide debris, where they nearly bury some of the outlying boulder piles; they interfinger northward with gravel brought from the hill slopes and from the Kashgan gorge. Jaidar Lake was about 9 km long, and it extended up the gorge for another 11 km, as shown by interbedded lake silts, river gravels,

Figure 1. Jaidar Lake silts with layers of colluvium, in gorge of Kashgan River.

Figure 2. Edge of cap of landslide debris (arrow) on crest of Kuh-i-Dufarush. Kabir Kuh is in the distance on the right.

JAIDAR LAKE SILTS AND SAIDMARREH LANDSLIDE

WATSON AND WRIGHT, PLATE 3
Geological Society of America Special Paper 123

colluvial deposits, and tributary fans (Pl. 3, fig. 1). Jaidar Lake's outlet was through the gypsum badlands around the northwestern end of the slide debris to Saidmarreh Lake, which was about 50 m lower in elevation.

In addition to the two major lakes, smaller ones were formed where valleys in gypsum were dammed. Several on the northwestern margins of the debris drained underground south to the Kashgan River. Chah Javal, the largest on the northeast, covered 1.3 km² (Fig. 2). It drained through the slide debris west to Jaidar Lake, which was about 100 m lower in elevation. A line of sinkholes marks the underground course for 2 km between the two basins.

Beyond the gorge through Kuh-i-Malah, the Kashgan River meanders across the Jaidar Lake floor and then cuts to the southwest 14 km through gypsum hills to the Saidmarreh Lake basin, where it joins the Saidmarreh River to form the Karkheh River. The Karkheh flows southeast 18 km through the landslide debris in the nearly straight Guardalanga gorge. The river has cut down to the gypsum beds beneath the debris and has carved two prominent terraces on the gypsum and the landslide debris (Pl. 1). The sloping terrace surface on the southern side of the gorge is about 75 to 100 m above the river, while the one on the northern side is about 45 to 60 m above the river. The gravel veneer on these surfaces mainly consists of stones of white limestone, firmly cemented. Rounded and angular stones are mixed; few are more than 15 cm in diameter, although some angular boulders exceed 1 m. The gravel is well-stratified and rests uncomformably on a surface beveled on dark-brown conglomerate or gypsum. The white gravel was derived from the slide debris, so it does not extend upstream from it. The range in elevation and the cuspate patterns of the terrace surface indicate that separate segments of each were cut not contemporaneously but during steady downcutting of the river. The two gross surfaces may represent two pauses in general downcutting, caused by large blocks in the debris dam or by major slumping along the gorge.

The dark-brown conglomerate forms the bed of the Karkheh River at several places. It dips in general conformity with the gentle anticlinal and synclinal folds of the gypsum beds on which it lies.

At several places along the Guardalanga gorge, beds of river-laid gravel 3 to 10 m thick are interbedded with landslide debris. Slumps involving 1,000,000 m³ of debris have occurred in the gorge, supplying the gravel and probably causing temporary damming of the Karkheh River.

COMPOSITION

Along the proximal margins of the slide debris on both sides of the Guardalanga gorge, the dull-white debris is composed of fragments of Eocene thin-bedded limestone, calcareous shale, and marlstone, weakly cemented in its upper part to form a breccia. The hills are rounded but steep-sided. The rock fragments are fissile, so talus is composed of flakes generally less than 15 cm in diameter. Larger blocks are present but are highly fractured and break into smaller pieces upon exposure. No blocks of Asmari Lime-stone occur in this area.

Northward this Eocene debris is overlain by a 10- to 30-m cap in which the same material is mixed with tabular blocks of buff Asmari Limestone up to 10 m across. Here cementation of the upper part of the mass is more thorough and may extend to a depth of as much as 30 m, so slopes are generally steeper and prominent buttresses have developed between talus chutes. Blocks protrude from the surface with random distribution (Pl. 4, fig. 1). Individual blocks are laced with semicircular solution channels (rillenstein) 1 to 10 cm wide in anastomosing patterns separated by sharp-edged ridges. Piles of sharp-edged chips up to 10 cm in diameter show the extensive solutional breakdown of blocks by the rain.

In some large exposures the slide mass appears from the distance to have a crude horizontal bedding. This could represent horizontal shear planes in the slide mass or subsequent slumping along steep slopes. The lineation may simply mark old, nearly obscured sheep trails, however. In the debris on a large hill west of the Mirabad lakes there are talus chutes up to 15 m wide and 100 m long on 45° to 60° slopes of breccia that largely mask the horizontal lines. The breccia breaks into fine fragments that are so widely dispersed that the talus cones at the base of the chutes are very small.

The rough superposition of Asmari blocks on top of Eocene debris preserves the original stratigraphic order on Kabir Kuh. Apparently the upper plate of Asmari Limestone slid forward on comminuted Eocene debris beneath (which in turn slid over the substratum), with relatively little tur-bulent mixing. The Asmari plate was broken into large blocks during trans-port; some of the blocks were carried forward to the leading edge of the slide, where there is a major concentration of them.

The slide debris is highly permeable, and most rainfall infiltrates. Several fresh-water springs along the southeastern margins of the landslide mass and along the Karkheh River flow either from the white gravel or over the top of the dark-brown conglomerate. Springs along the northwestern margins emerge from below the surface of level ground, so their sources are not apparent, but some may be related to buried gullies. A few of the closed depressions in the debris intersect the water table and thus contain spring-fed lakes. Two cold springs of brown, alkaline water occur near the eastern extremity of the landslide, one on each side of the Karkheh River. They bubble turbulently from round holes about 1 m in diameter in the mud. A crust of white salt lines the channels along which the water flows into the

Figure 1. Landslide scar on Kabir Kuh on skyline, with landslide debris in foreground. Mirabad lake in a graben. Note lack of large blocks on the graben escarpment (white on left and center) except at the top.

Figure 2. Landslide debris exposed in wall of sinkhole formed over gypsum. Note strong cementation of upper part of debris. Such small, round sinkholes occur only where the landslide debris is thin.

SINKHOLE AND SAIDMARREH LANDSLIDE

river. Their temperature, color, and turbulent flow suggest that they are points of resurgence for river water that has been captured upstream to flow underground through salt beds in the underlying gypsum.

ORIGIN

The immediate cause of the Saidmarreh landslide was probably an earthquake, for the region is seismically active. The problems concerning the localization of the slide and the mechanics of its movement and lubrication require more detailed consideration.

Harrison and Falcon (1938) suggest that a bulge or "knee-bend" slowly developed in the Asmari Limestone low on the flank of Kabir Kuh at the site of the future slide, because the weight of the slab was no longer supported from above after the crest of the anticline had been breached by erosion. The flexure of the bulge gradually increased as the Asmari plate continued to slip downslope on underlying thin-bedded limestones. Eventually, the flexure overturned and broke, providing the opportunity for the plate then to separate and slide to the base of the slope.

Such a surficial bulge on an anticlinal flank is also termed a "flap" by Harrison and Falcon (1934), who report its occurrence at numerous localities in the limestone folds of the Zagros Mountains. Direct evidence for a bulge in the area of the Saidmarreh landslide is lacking. Both east and west of the scar on the northern flank of Kabir Kuh the Asmari Limestone shows no bulge—only a uniform dip. In fact, at no point along the entire 160 km of Kabir Kuh does the limestone show a bulge, although two apparently occur along the southern slope of the Kuh-i-Dufarush anticline, where, in fact, the crest is not yet breached by erosion, implying that the features may have formed as a kind of subsurface drag fold during the tectonism rather than as a surficial structure. Of course, a bulge on Kabir Kuh may have been completely localized at the short segment in question, to be destroyed by the slide itself.

A second possible explanation for the localization of the slide is the undercutting of the Asmari Limestone at the northern base of the Kabir Kuh by the Saidmarreh River. Oberlander (1965) suggests that the Saidmarreh River may have been pushed against the base of the mountain by fans built by the Kashgan River, which probably entered the Saidmarreh in this area before being diverted by slide debris. Immediately downstream from the scar the Saidmarreh River now flows on Fars gypsum in the center of the synclinal valley between Kabir Kuh and Kuh-i-Dufarush, however, so there is no local evidence for undercutting. Farther east the Saidmarreh River does entrench the Asmari Limestone. Undercutting seems to be a better explanation than a flap structure for the localization of the slide.

The great distance of travel of the Saidmarreh landslide, as well as its climb 600 m over Kuh-i-Dufarush, required a low-friction slide surface. Harrison and Falcon (1938) postulate lubrication by water. Shreve, as a result of the study of the Blackhawk slide in southern California (1968), suggests that air-layer lubrication provided the low-friction surface for the Saidmarreh slide as well as for many other famous slides (1966). Kent (1966) suggests that air entrapment might explain the features of the Saidmarreh, Frank, and similar great slides. Because these suggestions for air lubrication in the Saidmarreh landslide are made without detailed reference to field relations, it is useful to examine here the criteria by which the hypothesis can be tested.

The Blackhawk slide of Shreve (1968), located in the desert area of southeastern California, started when a large mass of Blackhawk Mountain became detached. It moved 8 km down on the alluvial fans at the base of the mountain, with a total fall of the center of gravity of 600 m. Shreve proposes that the slide was launched from a bench near the base of the mountain, thereby entrapping air beneath it. The air, calculated to have formed a layer less than 30 cm thick, maintained a low-friction surface on which the debris moved, until upward escape of air through the slightly permeable slide mass reduced the air pressure and allowed the forward edge of the slide to sink to the ground. The obstruction thereby created at the front caused the formation of a frontal ridge and several wavelike ridges farther back. Some blocks of rock, intact during transport, were severely shattered by the impact as the slide halted, but the fragments were displaced only a few centimeters.

A launching platform, critical to the air-lubrication hypothesis, also existed in the case of the Tahoma Peak slide, which occurred on December 14, 1963. According to the reconstruction provided by Crandell and Fahnestock (1965), a mass of rock from the valley wall dropped 250 m to the glacier, gathered speed in sliding 6 km down the glacier (with a further vertical descent of almost 2000 m), acquired a basal air layer by being launched from the ice-cored terminal moraine (which stood 75 m above the foreground), and traveled an additional 3 km down the winding valley on a layer of entrapped air, caroming off first one valley wall and then the other. A terminal ridge is also present on the Tahoma slide.

After study of the landslide on the Sherman Glacier in Alaska in 1964, which he believes also was moved by air-layer lubrication (Shreve, 1966; Bull and Marangunic, 1967), Shreve compiled criteria for air-layer lubrication and listed five landslides that qualify by the presence of most of the features. The Saidmarreh landslide, which Shreve suggested is an air-layer lubrication feature, is not included in this list because of lack of information. It may now be examined with respect to these criteria:

(1) *Launching platform.* On the northern flank of the Kabir Kuh east and west of the slide, the 300-m-thick bed of Asmari Limestone forms a dip slope uniform to the base of the mountain, with no evidence of a structural or erosional bench from which a slide could be launched. The only possibility would be a terrace of the Saidmarreh River, but most of the terrace gravels in the region seem to be younger rather than older than the slide.

(2) *Local homogeneity of debris, indicating lack of large-scale turbulent mixing during sliding.* The Saidmarreh debris is marked by large blocks of Asmari Limestone on the surface, with smaller fragments from the older limestones and marls below, as seen in many deep exposures (Pl. 4, fig. 1). The composition does not vary laterally. The concentration of large blocks on the surface probably reflects the fact that the Asmari Limestone was above the other rocks when the slide began. Large-scale turbulent mixing therefore probably did not occur. Even if mixing had occurred, the large blocks could have worked their way to the top and floated on the dense mass, just as coarse particles are moved up and out in a mud flow (Hooke, 1967).

(3) *Wedge of rubble bulldozed to the distal edge of the slide, as a result of a sliding rather than a viscous motion on the substratum.* No such feature is found at the front of the Saidmarreh slide.

(4) *Lateral ridges, and* (5) *distal rim, formed when air escapes from the forward and lateral edges of the slide, causing debris to sink to the ground through its air cushion, to be piled up as a result of the momentum remaining in the rest of the slide mass.* The front of the Saidmarreh slide, where not buried by lake sediments, is steep and sharp, but the crest of the front is no higher than the general surface of the slide (Pl. 3, fig. 2).

(6) *Transverse tensional fissures, further indicating that motion was by sliding on a nearly frictionless substratum.* Portions of the network of ridges and troughs on the surface of the Saidmarreh landslide have a suggestive resemblance to some of the transverse fissures on the Sherman slide (Shreve, 1966). However, the much greater intricacy of the Saidmarreh pattern, combined with the slump character of the troughs, makes the alternative explanation of secondary slumping more probable.

(7) *Shattered but unseparated "jigsaw" blocks, formed when large blocks drop vertically as the entrapped air escapes, and the slide mass comes suddenly to a stop.* No such blocks are found in the Saidmarreh debris.

(8) *Debris cones of uncertain origin on boulders.* Such features are not found on the Saidmarreh debris, although the surface is so old that such minor features, if they ever existed, might not have survived the weathering and erosion of more than 10,000 years.

It can be concluded that there is little positive evidence that the Saidmarreh landslide mass was transported on a cushion of air. Another explanation must therefore be sought that will also explain the intricate pattern of surface features. The following sequence is proposed.

The slide was caused by an earthquake. The stage had been set by long-previous erosional breaching of the crest of the Kabir Kuh anticlinal ridge, and probably by the later undercutting of the 300-m-thick tilted plate of Asmari Limestone by the river at the northern base. The eastern and western margins of the slide plate were delimited by deep dip-slope gorges such as are now present on other segments of the hogback (Pl. 2).

The slide occurred rapidly, gaining enough momentum from the 900-m fall of its center of gravity to climb 600 m over Kuh-i-Dufarush to reach a point 18 km away. Shreve (1966) calculates a velocity of 330 km/hr, or more than twice as fast as most other famous slides, including those lubricated by air layers.

The lubrication for such a high velocity (if formation of an air-layer cushion was not possible) could have been provided by either or both of two conditions. First, the structural and topographic conditions are optimal for the entrance of water at the crest of Kabir Kuh and for its passage downdip under the cap of massive Asmari Limestone. The great hydrostatic pressure, caused by the 1300-m elevation difference between the ridge crest and the Saidmarreh River, may have increased pore-water pressure and thus reduced internal friction. There may also have been some interstratal solution between the Asmari Limestone and the underlying beds, thus decreasing the stability of the mass and increasing the water content. In any case, the thin-bedded Eocene marl under the Asmari cap may have been saturated with enough water to lubricate the comminuted marl at the base of the slide. Many land-slides can be directly related to temporary increase in the water content of potential slide planes. In other cases, however, no water increase is indicated— for example, the Hebgen Lake landslide in Yellowstone Park in 1959 occurred as a result of an earthquake after 6 weeks of dry weather (Hadley, 1964).

The second condition for lubrication is the substratum of gypsum over which the slide moved. Gypsum underlies the area from the base of the Kabir Kuh almost entirely across the Saidmarreh Valley to Kuh-i-Malah. Although the intricate badland topography provided a gross topographic roughness, the naturally smooth and slippery gypsum outcrops may have supplied a low-friction surface over which the slide could move, just as the Tahoma slide (Crandall and Fahnestock, 1965) moved 6 km down a slippery glacier surface.

The Saidmarreh landslide mass did not travel with turbulent motion throughout like a mudflow but instead slid forward either on its base or on the comminuted Eocene limestones. The Asmari Limestone plate remained on top, breaking and dispersing as blocks. This 15-km-broad plate, with some of the underlying Eocene beds on which it slid, moved off Kabir Kuh first and traveled farthest. As the plate left the mountain flank it drew along or was followed by a thick but narrower section of underlying Eocene beds, leaving a gash as deep as 300 m and 7 km wide in the center of the landslide scar (Pls. 1, 2). Much of this mass of lower beds moved only to the base of Kabir Kuh, where it is exposed south of the Guardalanga gorge with a relatively smooth surface without any capping of Asmari blocks.

MICROTOPOGRAPHY

Original microtopography is probably partially preserved as a series of faint troughs trending northeast parallel to the direction of sliding (pattern A). They are clearly crossed by, and are thus older than, the main pattern of microridges and troughs described next. The crossings convert the original features into rows of small depressions that form where the two sets of troughs cross (Pl. 5). The depressions hold more silt and water, so they support more vegetation and thus appear darker than the troughs on aerial photographs.

Elsewhere (Pls. 5–7) the slide is marked by an extensive series of small ridges and troughs (pattern B), generally with a northwest orientation, as well as by large grabens (pattern C) with the same orientation. The small troughs are commonly a few meters deep and up to 100 m wide and 2 km long. Sets of them commonly have a gentle arcuate trend. Their slopes appear to be stable. Many of the troughs are partially filled with silt. A reddish-brown soil has formed on the surface of the debris as well as in some of the silt fills.

Major grabens (pattern C) trend roughly parallel to bedrock structure. The largest, as much as 3 km wide and 100 m deep, extends for 15 km primarily along the axial projection of the buried, plunging Kuh-i-Dufarush anticline; it terminates in an abrupt rounded end 1 km before the northwestern edge of the slide (Pl. 6). Gypsum bedrock is exposed on its floor near the western end. The wide floor contains several horstlike splinters on which pattern B has survived. At the trough's western end, a sharp, transverse graben extends south to the Saidmarreh River on a straight line at right angles to the main trough.

Some grabens are actively forming now, so their scarps are light gray or almost white compared to the dark gray of the slopes on the stable patterns. Many grabens have been widened by slumping along the bounding scarps. Some of the lowered slump slices retain portions of pattern B (Pl. 5). Slices have moved down into the grabens without significant backward rotation, probably because the fractures extend beneath the graben floors, as do the main graben faults. Some grabens have been almost filled with such slices.

SURFACE FEATURES ON SAIDMARREH LANDSLIDE

Original flow features (pattern A) in exact center of photograph, oriented from upper left to lower right, intersected by minor ridges and troughs (pattern B), and these in turn by major grabens (pattern C) (*see* Plate 1 for location). Scale is about 1/27,000. North is to the base of the photograph.

MAJOR GRABENS AND OTHER SURFACE FEATURES ON
SAIDMARREH LANDSLIDE

Major grabens (pattern C) superimposed on the minor ridges and troughs (pattern B) (*see* Plate 1 for location). Badland topography on gypsum in the foreground is of the type overridden by the slide. Scale is about 1/63,000. North is to the base of the photograph.

WATSON AND WRIGHT, PLATE 6
Geological Society of America Special Paper 123

KUH-I-DUFARUSH AND SAIDMARREH LANDSLIDE

The plunging anticlinal Kuh-i-Dufarush (left) blocked the slide, but its nose was over-ridden (*see* Plate 1 for location). Most of the surface features on the slide belong to pattern B, but the lakes at top and bottom are in younger grabens (pattern C); note especially the sharp recent scarp (white) bounding the upper series. Note also the tiny sinkholes next to lakes of the upper series. The C^{14} date for the landslide comes from the sediment in the lake near the bottom edge of the photograph. Gypsum badlands are in lower left. Scale is about 1/63,000. North is to the base of the photograph.

WATSON AND WRIGHT, PLATE 7
Geological Society of America Special Paper 123

Others have been segmented into chains of small depressions, some of which contain ephemeral lakes (Pl. 5). The grabens transect and therefore postdate pattern B.

Active slumping is also occurring near the southeastern edge of the landslide debris, just north of the Karkheh River near the mouth of the Guardalanga gorge. Here a series of low scarps parallel to the river extend across the debris onto the white gravel terrace downstream. One scarp, 5 m high, within the slide continues into the terrace as a trough 5 to 10 m wide and 1 to 3 m deep. Also in this region of the slide debris is a recently formed graben that contains a string of six sinkhole lakes. The lakes are fed by underwater springs. Other springs emerge along the Guardalanga gorge at the top of the dark-brown conglomerate.

A set of three lakes occurs in the northwestern lobe of the debris west of the village of Mirabad (Pl. 4, fig. 1). These lakes are also fed by underwater springs. The largest is nearly 600 m long, 150 to 200 m wide, and at least 20 m deep. Each of the lakes has an old shoreline about 4 m above the present one; limestone boulders below this level are white and rounded by subaqueous solution, whereas boulders above this level are gray and are sharply etched by subaerial weathering. Two of the lakes were connected when the water level was highest. Six meters of marly sediment were cored from water two meters deep in the easternmost Mirabad lake. The basal sediment has a radiocarbon date of 10,370 $\pm$ 120 years B.P. (Y-1759). The graben in which the Mirabad lakes occur is probably much older than most of the other grabens, as indicated by the relatively stable appearance of its bounding scarps.

ORIGIN OF SURFACE FEATURES

Slide debris dammed the Saidmarreh and Kashgan Rivers and produced lakes covering large parts of their valleys. Saidmarreh Lake in the Saidmarreh Valley was at least 125 m deep initially, for its basin is filled with sediment to that thickness. Its outlet was east across the debris, but the channel probably did not erode rapidly because of a pavement of blocks. Harrison and Falcon (1938) imply that the lake overflowed at an elevation of 700 m and that it was stabilized at 685 m. Much lake water probably seeped through the dam of highly permeable landslide debris. Many small lakes in the landslide debris today have no surface outlet; the water seeps through the adjacent slide debris to the Karkheh River.

Jaidar Lake water, dammed in the Kashgan Valley to the north, probably also partly seeped through slide debris toward the Karkheh River and Saidmarreh Lake, for its elevation was 100 m above the Karkheh River near the slide mass and 45 m higher than Saidmarreh Lake.

The slide mass is limestone detritus; the underlying bedrock is largely gypsum. Both of these materials are subject to subsurface solution, especially by lake water seeping through the slide under high hydrostatic pressure. Subsurface flow could also remove the fine fractions of the debris, in part by piping (Wright, 1964). Subsequent roof collapse can account for the abundant small ridges and troughs (pattern B) on the debris surface. The intricacy of pattern B may reflect the variable thickness of slide debris over buried gypsum badlands. However, the slumping may have been shallow because Jaidar Lake water was seeping through the debris to maintain a high water table. Most of the flow, and thus the solution, takes place just at the surface of the water table and in a zone defined by its vertical fluctuation. Periodic changes in lake level (and thus ground-water level) would cause alternation of saturated and aerated conditions most favorable for solution and subsequent collapse.

The relatively stable aspect of pattern B implies that it is no longer actively forming. It is clearly being intersected by large grabens of pattern C, which result from deep-seated solution of gypsum rather than shallow solution and removal of slide debris. Pattern C formed after the big, slide-dammed lakes were drained by erosion of their outlets, and underground seepage was lowered into the more soluble gypsum beds beneath the slide debris. The large graben that follows the crest of the buried Kuh-i-Dufarush anticline may represent underground flow to the west down a now-buried gypsum valley that existed along the plunging axis of the anticline before the slide occurred. At the point where this crestal valley ended, the subsurface water escaped to the south down the anticlinal flank, producing the sharp graben that is transverse to the main one (Pl. 6).

Once the major grabens started to form, they continued to be widened by slumps along the flanks. The freshness of graben scarps and included slump blocks indicates that the process is still active. For example, some extremely delicate scarps can be found on a terrace of the Saidmarreh River that was cut into the slide debris after the dam was breached. Some of the grabens are old and inactive, however. For example, one of the Mirabad graben lakes in the northeastern lobe of the slide has basal sediment more than 10,000 years old, according to radiocarbon analysis.

Where the gypsum bedrock is exposed at the surface in the Saidmarreh valley beyond the limits of the slide, or where the slide debris is thin (Pl. 4, fig. 2), only a few small sinkholes are present, although the topography is quite irregular. The special conditions of abundant water flow from lakes under hydrostatic pressure through the debris and the gypsum underlying it produced extensive subsurface solution. The debris mantle, acting as a cap rock, confined the flow and thus allowed prolonged enlargement of underground channels by protecting them from immediate collapse. Collapse did not occur until the underground voids were large and semicontinuous. Without such a protective cap rock, solution and collapse in gypsum can occur, but the features are small and irregular as in that part of the Saidmarreh badlands beyond the limits of the slide.

Since its formation, the Saidmarreh landslide debris has experienced conditions similar to those that prevail in the wastage of debris-covered stagnant glacial ice. Subsurface melting of ice causes the formation of kettle holes, which widen by slumping along the margins, sometimes in slices that resemble those on the flanks of the grabens in the Saidmarreh debris (Pl. 8). The relative lack of linearity in most stagnant-ice topography reflects the lack of structural control in wastage of subsurface ice.

Another analogy is with the sinkholes and subsidence basins as much as 60 m deep in the Pecos Valley, New Mexico, formed by interstratal (subsurface) solution of Permian gypsum and limestone beneath a cap of gravel. Ground water comes from the Pecos River, which leaks underground during periods of low flow (National Resources Planning Board, 1942, p. 39–40—a reference supplied by J. F. Quinlan). Elsewhere in the Pecos Valley, as much as 1900 feet of alluvium have accumulated in depressions resulting from interstratal solution (Maley and Huffington, 1953).

Before this explanation for the surface features on the Saidmarreh landslide can be accepted, an alternative hypothesis must be considered. The patterns could be formed by lateral spreading off the buried crest of Kuh-i-Dufarush structure. The largest graben extends almost to the western end of the slide debris, where it terminates abruptly in a cross graben. The debris of the landslide, after coming to rest on top of the buried ridge, might have slid secondarily north and south off the crest, creating lateral tension that resulted in shear fractures at the crest, as well as one near the western end, resulting in the patterns described.

This hypothesis has several difficulties. Although the nose of the Asmari Limestone core of Kuh-i-Dufarush is buried, the ridge itself may not have extended far to the west. The plunge of the anticlinal structure, when projected westward, takes the resistant Asmari Limestone under the probable preslide badland surface of Fars gypsiferous red beds, as exposed on both sides of the unburied anticline as well as west of the debris. The anticlinal swell probably did not extend west of its limestone nose. It certainly has no topographic expression at the western margin of the debris, only about 1 km west of the end of the major graben, although still farther west the structure emerges again in a low limestone ridge.

Secondary slumping off the flanks of the buried ridge should produce a series of tensional faults with downthrow on the downhill side, rather than grabens, which imply vertical rather than lateral tension. In any case, the

Figure 1. Linear collapse depressions in till-mantled stagnant ice, St. Elias Range, southwestern Yukon. (Photograph by V. Rampton, Geological Survey of Canada.)

Figure 2. Collapse depression in till-mantled stagnant ice. Note slump slices on right. St. Elias Range, southwestern Yukon. (Photograph by V. Rampton, Geological Survey of Canada.)

COLLAPSE DEPRESSIONS ON STAGNANT GLACIAL ICE

structures cannot be viewed as a relaxation phenomenon produced immediately after emplacement of the slide, for the debris was apparently consolidated by cementation, at least in the upper part, before the grabens were formed.

We conclude, therefore, that linear patterns B and C, composed of troughs and aligned sets of depressions on the Saidmarreh landslide debris, were formed primarily by underground solution and subsequent collapse, resulting in features resembling those in areas of karst. The original slide was caused by an earthquake that released a mass of Asmari Limestone, which had been undercut by the Karkheh River. No evidence is found for the theory that the sliding mass had air-layer lubrication. The great speed, force, and range of travel of the Saidmarreh landslide debris was the simple result of an immense weight of rock released from sufficient height to slide catastrophically into a valley, over a smaller mountain, and beyond.

REFERENCES CITED

Bull, Colin, and Marangunic, Cedomir, 1967, The earthquake-induced slide on the Sherman Glacier, south-central Alaska, and its glaciological effects, p. 395–408 *in* Wura, H., *Editor*, Physics of snow and ice: Hokkaido Univ., Inst. Low Temp. Sci., Internat. Conf. Low Temp. Science (Sappora, Japan, 1966) Proc., v. 1, pt. 1, 711 p.

Crandell, D. R., and Fahnestock, R. K., 1965, Rockfalls and avalanches from Little Tahoma Peak on Mount Rainier, Washington: U.S. Geol. Survey Bull. 1221-A, p. A1–A30.

DeMorgan, Jacques, 1895, Mission scientifique en Perse: Etudes Geogr., v. 2, 331 p.

Hadley, J. B., 1964, Landslides and related phenomena accompanying the Hebgen Lake earthquake of August 17, 1959: U.S. Geol. Survey Prof. Paper 435Q, p. 107–138.

Harrison, J. V., and Falcon, N. L., 1934, Collapse structures: Geol. Mag., v. 71, p. 529–539.

—— 1937, The Saidmarreh landslip, southwest Iran: Geog. Jour., v. 89, p. 42–47.

—— 1938, An ancient landslip at Saidmarreh, in southwestern Iran: Jour. Geology, v. 46, p. 296–309.

Hooke, R. LeB., 1967, Processes on arid-region alluvial fans: Jour. Geology, v. 75, p. 438–460.

Jordan, P. G., 1956, Geology of Luristan: Bala Rud to Pul-i-Dukhtar: John Mowlem and Co., Iranian Roads Project Central Lab. Rept. 9008/1 (unpublished).

Kent, P. E., 1966, The transport mechanism in catastrophic rock falls: Jour. Geology, v. 74, p. 79–83.

Maley, V. C., and Huffington, R. M., 1953, Cenozoic fill and evaporite solution in the Delaware Basin, Texas and New Mexico: Geol. Soc. America Bull., v. 64, p. 539–546.

National Resources Planning Board, 1942, The Pecos River joint investigation: Washington, U.S. Govt. Printing Office, National Resources Planning Board.

Oberlander, Theodore, 1965, The Zagros streams: a new interpretation of transverse drainage in an orogenic zone: Syracuse Univ. Press, Syracuse Geography Ser., 168 p.

Shreve, R. L., 1966, Sherman landslide, Alaska: Science, v. 154, p. 1639–1643.

—— 1968, The Blackhawk landslide: Geol. Soc. America Spec. Paper 108, 47 p.

Wright, H. E., Jr., 1964, Origin of the lakes in the Chuska Mountains, northwestern New Mexico: Geol. Soc. America Bull., v. 75, p. 589–598.

GEOLOGICAL SOCIETY OF AMERICA, INC.
SPECIAL PAPER 123

Tree-Ring Dating of Snow Avalanche Tracks and the Geomorphic Activity of Avalanches, Northern Absaroka Mountains, Wyoming

NOEL POTTER, JR.
University of Minnesota, Minneapolis, Minnesota

ABSTRACT

Tree-ring dating in several avalanche tracks in Galena Creek valley, northern Absaroka Mountains, Wyoming, is used to determine the frequency of large snow avalanches that pass below the forest line. The following criteria are used: (1) datable scars on the trees, (2) changes in growth-ring pattern from concentric to eccentric, caused by tilting, (3) changes in growth rate due to increase in photosynthesis when adjacent trees are destroyed, and (4) age of trees within a given reforested avalanche track. The first two of these criteria are most reliable.

Many young trees within the avalanche tracks are protected by snow during avalanches, and they thus survive to reforest the track immediately following destruction of the larger trees.

Above the forest line, avalanche boulder tongues are one of the most reliable indicators of persistent activity in alpine regions, for they are formed over a period of many years by the accumulation of debris swept out of avalanche chutes. A peculiar linear feature on the surface of the tongues is the avalanche debris tail, which consists of fine debris that was deposited by large snow avalanches downslope 5 to 10 m from a large boulder. They are thought to be formed by a mechanism similar to that by which sand shadows are formed in the lee of obstacles in river channels or on desert dunes.

CONTENTS

INTRODUCTION

Snow avalanches have been a threat to man and his property ever since he first inhabited the mountainous regions of the world, but only in the past few decades has avalanche research become active. This work was pioneered in Switzerland, particularly at the Swiss Federal Institute for Snow and Avalanche Research at Davos. Avalanche research in the United States was initiated in the 1940's by the U.S. Forest Service, mainly because the threat to life was rapidly increasing as greater recreational use was made of the mountainous forest areas for skiing. Research on avalanche prediction and defense has advanced considerably in the past few years (U.S. Forest Service, 1961; Mellor, 1968), but little attention has been paid to the snow avalanche as a geomorphic agent (notable exceptions are Rapp, 1959, 1960, and Peev, 1966) and as an ecologic factor in forests. Some attention has been devoted in the

Alps to reforestation of avalanche tracks to stabilize snow in the starting zone (LaChapelle, 1966a).

This paper describes some of the distinctive depositional features of snow avalanches, considers the mechanisms by which the features are formed, and describes the effects of snow avalanches on forests just below timberline. The work was done in connection with a study of mechanisms of supply of debris and snow to rock glaciers in the northern Absaroka Mountains, Park County, Wyoming. Most of the work was carried out in the valley of Galena Creek (Fig. 1), a north-flowing tributary of Sunlight Creek. Elevations in the region range from 2380 m where Galena Creek joins Sunlight Creek to 3370 m along the divide of the Galena Creek drainage basin. The dominant bedrock of the region is Tertiary basaltic volcanics.

ACKNOWLEDGMENTS

I extend my appreciation to Dr. Herbert E. Wright, Jr., under whose supervision this work was carried out, for his encouragement and advice. Dr. Donald B. Lawrence made many helpful comments and suggestions concerning the ecologic aspects of this paper, and these are greatly appreciated. Robert Wood, James Meyers, Peter Waldo, and Noel Potter, Sr., served as field assistants and contributed significantly to the project. Richard Darling drafted the illustrations. Dr. John Montagne and Dr. Anders Rapp both visited me in the field and made many helpful suggestions concerning the effects of snow avalanches in the area. I am deeply indebted to Mr. and Mrs. Allan Weaver and Mr. and Mrs. Arthur Lantta of Red Lodge, Montana, who extended many courtesies to me during the project. I gratefully acknowledge the helpful comments of H. E. Wright, Jr., D. B. Lawrence, J. Montagne, C. L. Matsch, E. J. Cushing, R. LeB. Hooke, V. N. Rampton, and P. J. Conlon on various drafts of the manuscript.

This paper reports research undertaken in cooperation with the U.S. Army Natick (Massachusetts) Laboratories under Contract DA19-129-AMC-967 (N) and has been assigned No. TP-560 in their series of papers approved for publication. The findings in this report are not to be construed as an official Department of the Army position. The earlier phases of this work were supported in part by a grant from the Penrose Fund of the Geological Society of America.

CLIMATE

The climate of Galena Creek valley must be inferred from climatic data at nearby stations. This has been done (Potter, 1969, Ph.D. dissert., in prep.) with due cognizance of the notoriety of mountainous regions for extremely variable climates. The following information can give only a general indication of the climate in Galena Creek valley.

Mean annual temperature versus elevation was plotted for the 14 U.S. Weather Bureau stations within a 105-km radius of Galena Creek. The lapse rate apparent from the diagram is about 8.1° C per 1000 m of elevation. The plot suggests that the mean annual temperature in Galena Creek cirque at 3000 m elevation is at or below freezing. This conclusion is supported by the occurrence of permanently frozen ground at elevations as low as 2980 m on the mountains adjacent to Galena Creek. Permafrost has been reported from elevations as low as 2060 m at a site 64 km to the northwest in Yellowstone National Park (Good, 1964).

Precipitation in mountain regions is even more variable than temperature, partly because of the considerable redistribution of the winter's snow by wind drift. Mean annual precipitation at the nearest stations ranges from 48

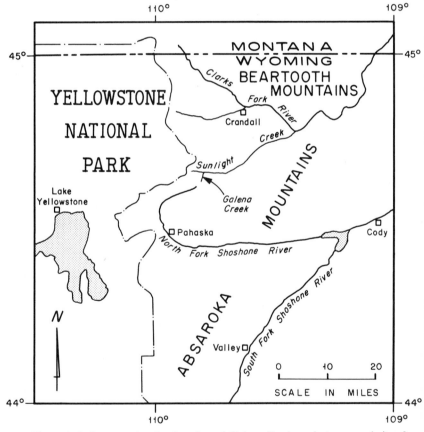

Figure 1. Index map showing location of Galena Creek project area and sites from which climatic data were obtained.

cm at Lake Yellowstone (elevation 2335 m, *see* Fig. 1) to 127 cm from a storage gauge at Valley (elevation 2707 m, *see* Fig. 1). Snow accumulates to depths greater than 6 m in Galena Creek cirque, but this almost certainly is a site of maximum wind accumulation in the lee of a major ridge. Snow accumulation in the forest at about 2550 m in the spring of 1966 was about 1 m, but the snowfall during the preceding winter was somewhat below average for the region. For example, at Sylvan Pass (elevation 2151 m), 34 km to the southwest, the 1938 to 1967 mean April 1 snow depth is 1.15 m, whereas the 1966 depth on April 1 was 77 percent of this mean (Peak and Crook, 1967).

Nearly all of the snow that falls in Galena Creek valley is blown about by strong winds. Knowledge of snow distribution is based on observations made during a flight over the area on February 25, 1967, and during work on the ground in May 1966. In general, north- and east-facing slopes are mantled by a heavy cover of snow, whereas west- and south-facing slopes have little snow cover. The snow is apparently blown from the west- and south-facing slopes by strong southwest winds, resulting in the formation of large snow cornices on the ridge crests facing north and northeast. These cornices persist into early July and are the source of many of the largest snow avalanches in the area.

VEGETATION AND THE FOREST LINE

The forest in Galena Creek valley is composed dominantly of whitebark pine (*Pinus albicaulis*) and subalpine fir (*Abies lasiocarpa*). At lower elevations, along Sunlight Creek, lodgepole pine (*Pinus contorta*) and Douglas fir (*Pseudotsuga menziesii*) are the common species, along with a few groves of aspen (*Populus tremuloides*) that grow immediately adjacent to the river. The forest line (upper edge of continuous forest) in Galena Creek valley lies between 2620 m and 2850 m. Within an 8-km radius of Galena Creek, the forest line has the following mean elevations with respect to slope aspect:

north-facing slopes	2880 m
east-facing slopes	2820 m
south-facing slopes	2580 m
west-facing slopes	2970 m

The low elevation of the forest line on south-facing slopes is particularly accentuated in Sunlight Creek valley (Pl. 1, fig. 1), where on opposite sides of the valley the elevation difference of the forest line is as great as 360 m. In this case, bedrock is exposed almost everywhere down to the forest line on the south-facing slopes, whereas the north-facing slopes of the valley have a thick mantle of colluvium in which trees can take root.

Krummholz, a narrow zone of stunted, contorted trees, is usually present above the forest line. In most places even alpine vegetation of grasses and

Figure 1. Aerial photograph of Galena Creek project area. North is toward bottom of photograph. Sunlight Creek flows eastward (left) across the bottom; Galena Creek flows northward in the center. Terminus of Galena Creek rock glacier is at A. Typical non-forested avalanche track is at B. Seven forested tracks are labelled C, the right-hand one of which is enlarged in Plate 2, figures 1–3. U.S. Forest Service 1933 Aerial Photograph No. 947.

Figure 2. Trees at forest line showing branches at base of trunks that were protected by winter snow, along with trimmed branches just above.

GALENA CREEK AREA AND TREES TRIMMED BY AVALANCHES

POTTER, JR., PLATE 1
Geological Society of America Special Paper 123

sedges is sparse, because of the coarseness of the talus, steepness of the slopes, and abundance of bedrock exposures. A few small patches of turf occur on flat or gently sloping mountain tops, but for the most part the ridge crests are sharp.

VEGETATIONAL FEATURES OF AVALANCHE TRACKS

Types of Avalanche Tracks

One of the most striking features of the forests of the region is the abundance of fingerlike avalanche tracks (Fig. 2 and Pl. 1, fig. 1) that reach from the forest line down to elevations as low as 2340 m. These tracks are typically narrow at the top and expand in width downslope. Most of them head in gullies or chutes that extend downward from the mountain crests. These tracks are of two types.

Nonforested Tracks. Nonforested tracks contain almost no growing trees and are covered only by grass and sedge turf. In general, they are light-colored on aerial photographs (Pl. 1, fig. 1). They apparently represent sites where avalanches occur frequently enough to keep all but the very youngest or shortest seedlings cleared from the tracks.

Every summer from 1963 to 1966, debris and broken tree limbs were observed on top of snow at the base of most of the chutes and tracks that extend to the bottom of Sunlight Creek valley. Avalanches must run down these nonforested paths nearly every winter. The few trees that do exist in these tracks are small—usually less than about 1 to 3 m high. The trees are taller and more abundant in depressions than on high spots. Observations during May 1966 suggest that most of the small trees in the nonforested tracks are partially buried by snow during winter and protected from destruction by avalanches. The tops of most of these trees, however, show evidence of trimming by avalanches. Thus the maximum height of trees should approximate the winter snow depth at that locality. Trees that grow in depressions where snow accumulates to greater depths grow to a greater height.

The position of scars and trimmed branches on trees at the forest line shows the same relations (Pl. 1, fig. 2). On the downslope side of the trees there are numerous branches, but on the upslope side the branches are confined to the high part of the trunk and a narrow zone at the base, where they are protected by fallen and blown snow from small avalanches or rockfalls. In depressions, branches occur higher above ground (up to 2 m) than on low ridges, so an imaginary plane through the scars on the trees has a uniform slope relative to the irregular microtopography. Drifting of snow in winter fills in swales and smooths out the topography, and the scarred portion of the tree trunks is produced by avalanches that run on top of the snow. The presence of stumps from which the upper portion of trees have been sheared off at the same level as the scars on the living trees suggests that avalanches rather than rockfalls are responsible.

Some of the nonforested tracks contain well-developed stream channels, often with debris-flow levees, whereas approximately an equal number have no well-defined channels. Water may be important in maintaining some of the tracks, but the ones without channels are treeless as a result of avalanches only.

Forested Tracks. Forested tracks are covered with young living trees. In particular, forested tracks are outlined by distinct trim lines that separate older forest from the younger growth in the path. They are dark-colored on aerial photographs, except that immediately after forest destruction by an avalanche they appear as an intermediate shade of gray, becoming gradually darker as new trees fill in the path (Pl. 1, fig. 1, and Pl. 2, figs. 1–3). They do not generally extend as far downslope below the forest line as do the non-forested tracks. They are commonly covered with trees that have been up-rooted or broken. The trees are almost always aligned parallel to one another with the tops pointing downslope.

Successive avalanches may be recorded by a succession of trim lines. The frequency of large snow avalanches that run below the forest line can be determined by tree-ring dating of trim lines, as has been done with recent glacier advances (Lawrence, 1950; Sigafoos and Hendricks, 1961), volcanic events (Druce, 1966; Lawrence, 1941, 1954), rock avalanches (Heath, 1960), and floods (Sigafoos, 1964). Trim lines of individual avalanches are not often found adjacent to nonforested tracks, probably because the avalanches responsible for these run nearly the same course year after year.

Galena Creek Tracks

The avalanche tracks studied are shown in Figure 2 (*see also* Pl. 1, fig. 1). One of these tracks, exhibiting several trim lines, was selected for detailed study (Fig. 3 and Pl. 2, fig. 4). It is located just downvalley from the terminus of Galena Creek rock glacier. This Galena Creek track occurs in forest composed of subalpine fir and whitebark pine. The three trim lines in this track have been dated by standard tree-ring methods and, in the case of the 1963 and 1965 avalanches, by ground observations and photographs.

In a sample of 50 of the trees destroyed by the 1963 avalanche, 43 were subalpine fir and 7 were whitebark pine. The firs ranged up to 13 m in height and 28 cm in basal diameter, and the pines to 8.5 m in height and 25 cm in basal diameter. Of the firs, 29 were uprooted and 14 were broken off either just above the base or up to 2 m above the base. All 7 of the pines were up-rooted.

Most of the tree tops point in the direction of travel of the avalanche, but at least 14 pointed in the *opposite* direction: 8 of these 14 were broken off

Figure 1

Figure 2

Figure 3

Figure 4

Figure 5

AVALANCHE TRACKS IN SUNLIGHT CREEK AND GALENA CREEK VALLEYS

Figures 1–3. Sequence of photographs showing progressive healing of an avalanche track on the southern side (north is to right in photographs) of Sunlight Creek approximately 2.4 km west of confluence with Galena Creek. Age of avalanche that produced track is unknown. Figure 1: 1933, from U.S. Forest Service Photograph No. 947; figure 2: 1949, from U.S. Geological Survey Photograph GS-JB-3-113; figure 3: 1958, from U.S. Forest Service Photograph ECF-14-101.

Figure 4. Galena Creek avalanche track, looking east. Prominent trim line (dashed) is the result of an 1884 avalanche. Recent track in the middle is from a 1963 avalanche. *Compare* Figure 3.

Figure 5. Trees destroyed by 1963 avalanche in Galena Creek avalanche track. Note small trees still standing that were not destroyed by the avalanche.

POTTER, JR., PLATE 2
Geological Society of America Special Paper 123

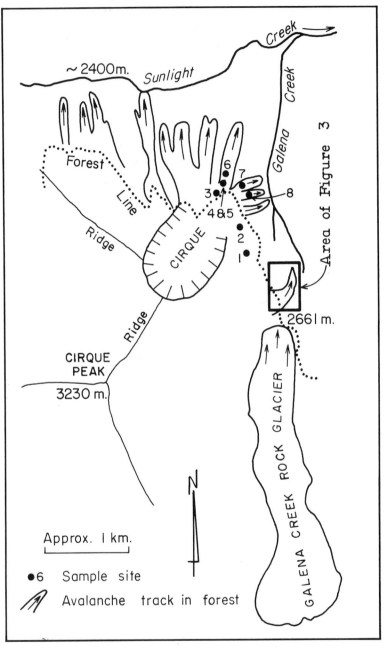

Figure 2. Location of avalanche tracks in and adjacent to Galena Creek valley and sites from which samples were collected for tree-ring analysis.

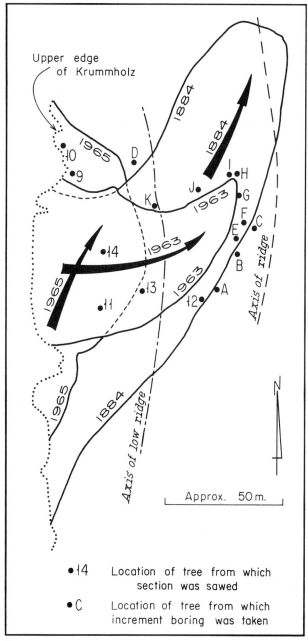

Figure 3. Galena Creek avalanche track, showing location of samples collected for tree-ring analysis.

at the base and 6 were uprooted. Ten of the fourteen were also broken off part way up the trunk at heights ranging between 4 and 8.5 m from the base. One might expect that most of the trees destroyed by an avalanche would be tipped in the direction of travel of the avalanche, but an explanation should be found for the disoriented trees. This problem will be reserved for discussion in a later section.

In 1966, numerous small trees were still growing in the tracks of the 1963 and 1965 avalanches (Pl. 2, fig. 5). The trees range in height from 1 to 3 m; despite their relatively small diameters (up to 10 cm), some are more than 100 years old. In forested tracks, such trees are ready to reforest the path immediately. These trees are much scarred and trimmed of branches and tops, but they are small and flexible enough to withstand the passage of the avalanche without being killed. Some of the smaller trees become partially uprooted and tilted by the avalanche but still manage to survive. Many are sheared off 1 to 2 m above the base, having been protected below by snow.

The small trees that survive were used to date the older avalanche tracks. Trim lines adjacent to the tracks were dated with the use of the following criteria: (1) scars (Pl. 3, fig. 1), although scars found on single trees might result from rockfalls across the winter snow surface, (2) change from concentric to eccentric growth due to tilting (Pl. 3, fig. 2), (3) maximum absolute age of the younger group of trees within a specific track; even though some of the small trees are not destroyed, many new seedlings can start up when the large trees are destroyed, and if enough young trees are sampled, the maximum age of this group indicates the approximate date of the avalanche, and (4) either abrupt increase in growth rate in the older trees adjacent to the track resulting from the loss of competition provided by the many larger trees destroyed by the avalanche, or abrupt decrease in growth rate resulting from injury to the tree or resulting from increased drying caused by removal of shade furnished by the adjacent trees that were destroyed (Pl. 3, fig. 3). Such changes in growth rate might be found adjacent to a track along the trim line, as well as within the track.

———————————→

TREE SECTIONS SHOWING AVALANCHE DISTURBANCE
Small divisions on scale are millimeters.

Figure 1. Section 1-U, showing scars at 1918 (A), 1939 (B), and 1945 (C).

Figure 2. Section 2-U, showing change from concentric growth and decreased growth rate at 1911.

Figure 3. Section 14-L, showing increased growth rate along some radii and decreased growth rate along others following 1963 avalanche in Galena Creek track.

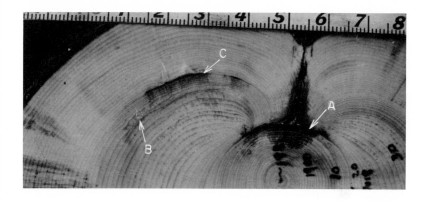

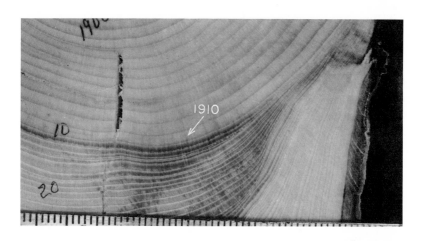

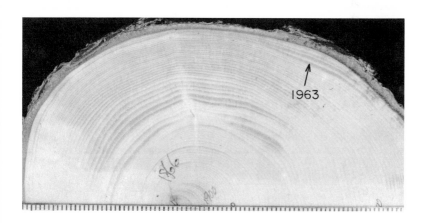

POTTER, JR., PLATE 3
Geological Society of America Special Paper 123

The causes of abrupt changes in growth rate are not always easy to interpret. Several cases were found in which, following the 1963 avalanche, trees that were not even apparently tilted or scarred grew more rapidly along some radii and more slowly along others (Pl. 3, fig. 3). Abrupt changes in growth rate were accepted as the result of disturbance by an avalanche if corroborative evidence in the form of scars or increased eccentricity could be found for the same year in other trees.

It should be apparent from the illustrations of typical tree sections (Pl. 3) that different radii along which increment borings might have been taken could yield different information. For this reason, sawed sections give a much more reliable record of disturbance. Much of the record can be obtained from trees that have been killed by avalanches of known date, so damage to the forest by sampling is not great.

Figure 4 shows in diagrammatic form the data for both the sections and the increment borings from the Galena Creek track. The 1884 avalanche that formed the prominent trim line in Plate 2, figure 4, is recorded in almost all sections. Increment borings E through K are from trees within the 1884 path. These trees started to grow 6 to 13 years after the 1884 avalanche; probably they represent new seedlings. Sections 11, 13, and 14 were apparently young trees not destroyed by the 1884 avalanche; they record the avalanche by an increase in growth rate and by increased eccentricity due to tilting. Numerous decaying tree trunks cover the forest floor within the 1884 path, oriented with their tops pointing in the direction of travel of the avalanche.

An event is recorded in several of the samples for the year 1935, but no trim line could be found that corresponds to this event. If a trim line was formed in 1935, it may have been destroyed in 1963. Perhaps the avalanche responsible for the 1935 scars and changes in growth rate was not large enough to produce a trim line but only to selectively destroy weak trees and scar and disturb a limited number.

The 1963 and 1965 avalanches are manifest in the sections by abrupt changes in growth rate and were marked by fresh scars on some of the trees.

Several changes in growth rate and one case of increased eccentricity were found that do not correspond to the reasonably well-documented events of 1884, 1935, 1963, and 1965. Although these might represent disturbance by small avalanches, they should not have much significance attached to them, for there are many factors that might effect an abrupt change in growth rate of a single tree.

Other Tracks

Certain years have had exceptional avalanche activity because of favorable weather conditions. Several other tracks (Fig. 2) were sampled to determine whether or not a record of avalanches corresponding to the years represented in the Galena Creek track could be found. The record from these other tracks, here called the Cirque Peak avalanche tracks, is shown in Figure 5. Only one of these tracks was sampled by sectioning more than one tree. This track is represented by Sections 3, 4, 5, and 6 (Fig. 5). The only event recorded in more than one of these four sections is one for 1956. There appears to be little correspondence of dates from track to track, and no correspondence of events with those in the Galena Creek tracks. It is thus concluded that, at least locally, there is no evidence to support the suggestion

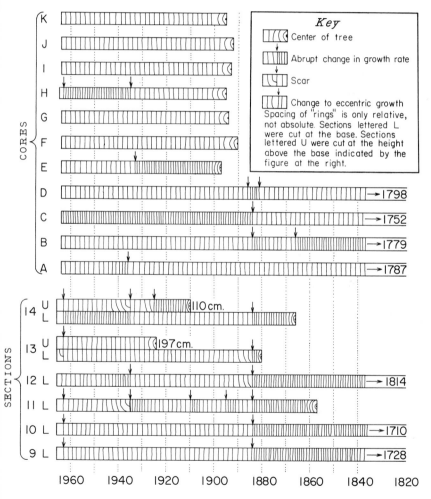

Figure 4. Tree-ring cores and sections from Galena Creek avalanche track. For sample site locations *see* Figure 3.

that major destructive avalanches run into the forest in many places the same year.

GEOMORPHIC FEATURES OF AVALANCHES

The most distinctive geomorphic feature of high altitudes in the Absaroka Mountains, aside from the rock glaciers (Potter, 1968), is the abundance of well-developed avalanche boulder tongues (Pl. 4, fig. 1), such as those first described by Rapp (1959) from the mountains of Swedish Lapland. The tongues are composed of debris that is swept out of avalanche chutes and

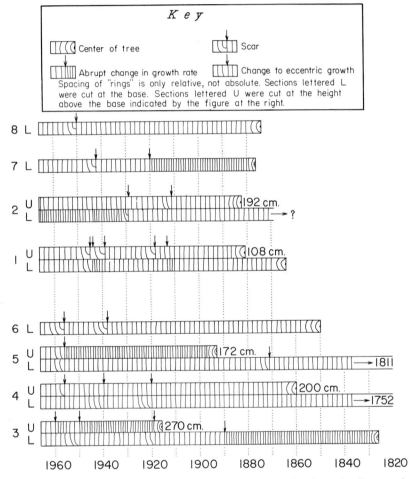

Figure 5. Tree-ring sections from several Cirque Peak avalanche tracks. For sample site locations *see* Figure 2.

Figure 1

Figure 2

Figure 3

AVALANCHE BOULDER TONGUES AND DEBRIS TAILS IN GALENA CREEK

Figure 1. Aerial photograph of avalanche boulder tongues in Silvertip Basin at the head of the North Fork of the Shoshone River, 2 km south of the head of Galena Creek valley. North is toward the lower-left corner of the photograph. Downvalley is toward the right (southwest). The long dimension of the photograph represents approximately 1.7 km on the ground. Maximum relief within the area of the photograph is about 300 m. Note the downvalley deflection of several tongues. U.S. Forest Service 1933 Aerial Photograph No. 944.

Figure 2. Avalanche debris tail on Galena Creek avalanche boulder tongue, looking downslope.

Figure 3. Source and upper part of avalanche boulder tongue (B) in Figure 6. The avalanches that feed the tongue have cut a swath through the krummholz where the roads cross the tongue. Note alluvial cone (A) at base of chute.

POTTER, JR., PLATE 4
Geological Society of America Special Paper 123

deposited in tonguelike form by many successive avalanches. They are found almost exclusively above the forest line, which suggests that the frequency of avalanches there is much higher than that of avalanches that pass into the forest below. Many of the avalanche tongues have probably been building up since the Pleistocene, because what little evidence is available suggests that they accumulate very slowly and because, in some cases, they cover Pleistocene lateral moraines.

Rapp (1959, p. 35) distinguishes two types of avalanche boulder tongue: (1) road-bank tongues which are raised like flat-topped road banks above the surrounding surface on which they are deposited, and (2) fan tongues, which are composed of a thin, fan-shaped cover of debris. Most of the tongues of the northern Absaroka Mountains are of the road-bank type, although I have seen fan tongues in the Beartooth Mountains near the Montana-Wyoming state line. Avalanche boulder tongues have previously been reported from the United States only in a cursory manner (White, 1968).

Typical road-bank tongues in the northern Absaroka Mountains are 50 m wide and extend downslope as much as 200 m from the base of the bedrock chute from which they originate. They may stand as much as 3 m above the surface on which they are deposited. In some cases, immediately downslope from the avalanche chute, there is an alluvial cone, composed of debris washed from the chute during spring snowmelt and during summer cloudbursts. Slopes on these cones are typically 20° to 25°. The debris on the cone commonly has a smaller modal grain size than that on the tongue below. The cone merges downslope with the avalanche tongue. The tongue commonly has a longitudinal slope less than about 15°, and in a few cases tongues cross the valley floor and rise up the opposite side of the valley.

Avalanche boulder tongues often have peculiar linear features on their surfaces called avalanche debris tails (Pl. 4, fig. 2). These features, also first described by Rapp (1959), are considered by him to be one of the most characteristic features of avalanche depositional areas and are found almost exclusively above the forest line. The debris tails usually have a large boulder (30 to 200 cm in diameter) at the head and a "tail" of finer debris (1 to 20 cm in diameter) that starts nearly as high as the boulder and then tapers to the level of the tongue 5 to 10 m downslope. There is usually a pile of debris on the uphill side of the boulder at the head, but this is much shorter and blunter than the tail. In some cases, below an exceptionally large boulder, a double tail is found. Rapp (1959, p. 40) suggested two possible mechanisms for the formation of debris tails. (1) The fixed boulder halts stones that collide against it when they travel in the base of an avalanche. These stones hit the boulder and roll over it, to be dropped (and collected) in the lee of the large boulder. (2) Fine loose debris that was previously deposited by dirty avalanches is protected from removal by avalanche snow by the fixed boulder.

An avalanche tongue and its associated debris tails were investigated in Galena Creek just below the rock glacier (Fig. 2 and Pl. 4, fig. 3). A topographic map of the tongue shows the orientation of debris tails, wood fragments, and rock fragments on its surface (Fig. 6).

Three burlap carpets placed across the largest debris tail on the Galena Creek tongue (Pl. 4, fig. 2) in 1964 collected several rock fragments in the winter of 1965–1966. One of the rocks holding down the corner of a burlap sack was rolled downslope halfway across the sack, but this rock had been placed beside the tail rather than upon it. The evidence collected in this experiment suggests (1) that debris is presently being deposited on the debris tail, although perhaps not preferentially compared to the rest of the avalanche boulder tongue surface, and (2) that rock fragments projecting above the tongue surface may be moved short distances by individual avalanches. The problem of debris tails will be considered further in the discussion at the end of this paper.

The Galena Creek avalanche tongue is asymmetrical in transverse profile, as are many of the other tongues in the region (see Fig. 6). Rapp (1959, p. 41) suggested that the snow deposited in the lee of a tongue by strong winter winds protects that side and deflects avalanches to the windward side, where debris builds a gentler slope. On the Galena Creek tongue snow accumulates to a greater depth on the southern side because of northerly winter winds, and it remains here longer into the summer despite more exposure to the sun. The debris tails on the northern side trend northeast, whereas those on the central and southern part of the tongue are parallel to its axis (Fig. 6). The same is true for the long axes of debris fragments. This evidence appears to support Rapp's hypothesis for the deflection of avalanches and the development of the asymmetry.

Although the avalanche tongues are mainly constructional landforms, there is evidence that erosion also occurs on the tongue. Rapp has suggested (1965, personal commun.) that coarse fragments project far enough above the surface of the tongue that they are swept along the surface by successive avalanches. Perhaps the avalanches that carry debris onto the tongue are occasionally followed by cleaner, high-velocity avalanches that sweep the larger particles off the surface. The net result of this process should be a fringe of coarse debris that encircles the tongue. This indeed seems to be the case for the Galena Creek tongue (Fig. 6). In both transverse and longitudinal profile, the modal size of debris at the edges of the tongue is much coarser than that in the central part.

The debris sizes shown in the figure represent the intermediate diameters of surface fragments, which were determined with a variation of the method developed by Wolman (1954) for sampling coarse gravel in a river bed. The technique used in this case was to place a surveying rod horizontally on the

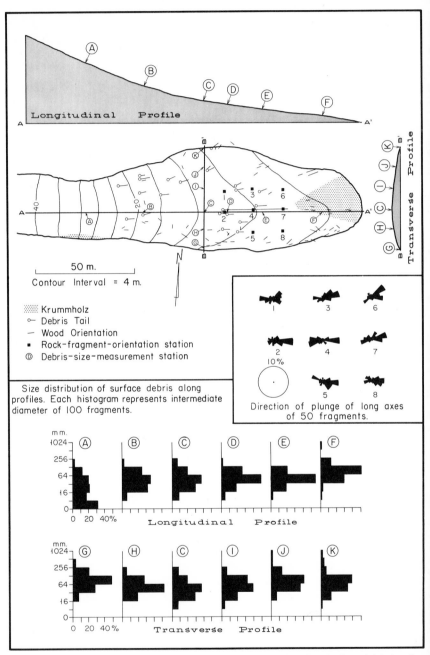

Figure 6. Galena Creek avalanche boulder tongue, showing distribution of debris tails, fragment orientation, and debris size distribution.

tongue. Samples were taken directly beneath marks on the surveying rod at regular intervals larger than the largest fragments in the immediate area. Each fragment was measured and replaced. When the full length of the rod had been traversed, the rod was moved laterally a distance equal to the spacing between samples along the rod. Sampling was continued until 100 rocks had been measured.

The origin of the avalanche debris tails remains enigmatic. It is clear that the erosional mechanism of origin, in which the large boulder at the head would protect the debris in the tail from being eroded away, is not satisfactory because it does not explain why double tails are sometimes found. The depositional mechanism of origin seems more reasonable. Rapp (1959, p. 40) points out the similarity of the form of avalanche debris tails to crag-and-tail moraines. Also similar are sand shadows formed downwind from obstacles in desert dune environments (Bagnold, 1941, p. 189–191) and sand shadows formed downstream from obstacles in aqueous environments (Pettijohn and Potter, 1964, p. 335; Potter and Pettijohn, 1963, p. 121 and Plate 15A).

Bagnold's (1941) description of the mechanism of formation of sand shadows by wind is instructive. In the incipient stages, two wings, or tails, are formed after sand piles up on the windward side of the obstacle and rolls out into the stream of sand that is already passing the sides of the obstacle. The two tails gradually coalesce to form a rather diffuse shadow that is wider than the obstacle behind which it forms. But Bagnold (p. 191) points out that usually the shadow's shape is modified to a tapered form that narrows away from the obstacle due to slightly shifting wind directions that allow the sides of the shadow to be "trimmed" away. It is clear in the case of sand shadows that deposition occurs in the lee of the obstacle because eddies occur there and the forward wind velocity is less than that outside the wind shadow.

Perhaps, as Rapp (1959) suggests for his depositional model, rock fragments in the base of avalanches bounce off the boulder at the head of the debris tail and come to rest in the "low-velocity zone" in the lee of the boulder. Perhaps also the tails are "trimmed" by successive avalanches so that the distal ends are kept tapered rather than diffuse. The length of the tails (as much as 10 m) implies a high degree of constancy of the travel direction of successive avalanches. The same is implied by the consistent trend of the tails in a parallel or fan-shaped pattern on individual avalanche tongues. It is clear that the tails must be formed where there is little snow accumulation in winter or else the avalanches that form the tails must scour through any accumulated snow to the tongue surface. This is implied as well by the coarse debris that is apparently swept to the fringes of the avalanche tongues. Limited observations in Galena Creek suggest that little snow accumulates on the avalanche tongue, because it is blown off the ridges into adjacent hollows.

DISCUSSION

LaChapelle (1966b) has applied statistics to the problem of estimating encounter probabilities for avalanches in a given avalanche track. He points out that the greatest difficulty in applying the tables that he has developed to the problem of evaluating avalanche hazard is that of determining the return interval of avalanches in a given track. He suggests a minimum possible single return interval can be determined from tree growth. With the techniques applied in the Galena Creek area, and given favorable conditions in the form of trim lines or sufficient trees in the avalanche track, it should be possible to obtain more than one return interval for a given track. Of course historical records of avalanche frequency would be much better than tree-ring evidence, but these records are available for relatively few localities in North America, and for only 20 to 30 years in most cases (LaChapelle, 1966b).

In an area such as the Absaroka Mountains, where avalanches commonly pass into the forest from the ridges above, an opportunity is available to examine many of the unanswered questions concerning the ecologic effects and geomorphic activity of snow avalanches. A systematic program of dating avalanche paths over a larger area than that considered here could be undertaken to determine the frequency of large snow avalanches and the relation of these to the climatic record, albeit short, for the region. It is possible that statistical analysis of tree-ring widths for some of the species growing near the forest line could give some of the needed climatic information for comparison with avalanche frequencies. Fritts (1965, 1966), for example, has considered the relations between climate and tree-ring width in western North America and has prepared maps of high and low moisture for the period since 1500 at 10-year intervals.

The forested tracks, which are recolonized relatively rapidly, raise the question of the effects of snow avalanches on the composition of the forest near the forest line. It should be determined whether the new growth of trees in the avalanche tracks is dominated by certain species to the exclusion or suppression of others in the surrounding forest, or whether the composition of the forest in the tracks is essentially the same as that of the surrounding forest.

Unfortunately, the types of avalanches responsible for the tracks below the forest line and for the avalanche tongues can only be inferred. There are many bases for avalanche classification (U.S. Forest Service, 1961, p. 17–28; Mellor, 1968, p. 17–26; DeQuervain, 1966, p. 413). The most useful criteria (DeQuervain, 1966) for this discussion are the position of the sliding surface and the form of movement.

It has already been noted that the avalanches that form the tongues and the associated debris tails must be full-depth avalanches by the time they reach the tongues. The avalanches that pass into the forest may be either full-depth or surface-layer avalanches, depending in large part upon the depth of accumulated snow over which the avalanche passes. In many of the forested paths, trees broken as much as 2 m above the ground, as well as surviving trees with branches preserved at the base but trimmed away at higher levels, suggest that at least locally the avalanches move upon the surface of the snow previously accumulated in the forest.

Airborne avalanches in general move farther out across gentle slopes and have much higher velocities than do flowing avalanches. By both of these criteria airborne-powder avalanches are favored as the primary type responsible for the tracks and tongues. The air blast that is commonly associated with airborne-powder avalanches (Mellor, 1968, p. 73) might most easily accomplish the toppling and breaking of a large number of trees in the forested tracks. The Galena Creek paths lie beneath a precipitous cliff that is about 300 m high; above this the slope is gentler to the ridge crest. If avalanches originate near the ridge crest, the drop over the cliff would almost certainly result in an airborne avalanche.

One might expect that most of the trees destroyed by an avalanche would be toppled in the direction of travel of the avalanche. It was noted earlier that some of the trees in the Galena Creek path were disoriented, with their tops pointing in the direction from which the avalanche came. Perhaps the wind blast that accompained the avalanche extended only 6 to 9 m above the ground and swept the base of a few of the trees forward ahead of the tops of the trees. This might explain the breakage of these trees partway up the trunk. It may also be significant that most of the disoriented trees were found at the distal end of the track.

Rapp (1959) has suggested that avalanche boulder tongues could, under the appropriate conditions of occurrence, preserve a unique record of the importance of snow avalanches as a denudational agent. A quantitative estimate of the amount of debris removed from chutes per unit time by avalanches could be made if the age of the surface on which the avalanche tongues have accumulated is known. Unfortunately, despite the development of the skeleton of a Neoglacial chronology for the North American Cordillera (Porter and Denton, 1967), the lack of datable material from most of the Rocky Mountain ranges, including the Absarokas, makes it difficult to assign an absolute age to most of the glacial deposits.

The Galena Creek avalanche tongue and several adjacent ones are deposited on top of an inactive portion of the Galena Creek rock glacier. These tongues are undeformed and thus they must have accumulated since deactivation of this portion of the rock glacier. Other tongues, which have accumulated on an upvalley, active portion of the rock glacier, are deformed in the downvalley direction by the movement of the glacier. If it were possible to date these portions of the rock glacier, an excellent opportunity to determine the rate of accumulation of the avalanche tongues would be available.

In conclusion, I believe that the problems of the effects of snow avalanches on forest ecology at the forest line and the role of snow avalanches as a geomorphic agent of denudation deserve much more attention in North America than they have heretofore received.

REFERENCES CITED

Bagnold, R. A., 1941, The physics of blown sand and desert dunes: New York, William Morrow and Company, 265 p.

DeQuervain, M. R., 1966, On avalanche classification, a further contribution, p. 410–417 *in* International symposium on scientific aspects of snow and ice avalanches (Davos, 1965): Internat. Assoc. Sci. Hydrology Pub. no. 69, 424 p.

Druce, A. P., 1966, Tree-ring dating of recent volcanic ash and lapilli, Mt. Egmont: New Zealand Jour. Botany, v. 4, p. 1–41.

Fritts, H. C., 1965, Tree-ring evidence for climatic changes in Western North America: Monthly Weather Rev., v. 93, p. 421–443.

—— 1966, Growth-rings of trees: their correlation with climate: Science, v. 154, p. 973–979.

Good, J. M., 1964, Relict Wisconsin ice in Yellowstone National Park, Wyoming: Wyoming Univ. Contr. Geology, v. 3, p. 92–93.

Heath, J. P., 1960, Repeated avalanches at Chaos Jumbles, Lassen Volcanic National Park: Am. Jour. Sci., v. 258, p. 744–751.

LaChapelle, E. R., 1966a, Reforestation in the Austrian Alps: Alta, Utah, Alta Avalanche Study Center Misc. Rept. no. 9, 16 p.

—— 1966b, Encounter probabilities for avalanche damage: Alta, Utah, Alta Avalanche Study Center Misc. Rept. no. 10, 10 p.

Lawrence, D. B., 1941, The "floating island" lava flow of Mt. St. Helens: Mazama, v. 23, p. 56–60.

—— 1950, Estimating dates of recent glacier advances and recession rates by studying tree growth layers: Am. Geophys. Union Trans., v. 31, p. 243–248.

—— 1954, Diagrammatic history of the northeast slope of Mt. St. Helens, Washington: Mazama, v. 36, p. 41–44.

Mellor, M., 1968, Avalanches: pt. 3, sec. A3d, 215 p. *in* Sanger, F. J., *Editor*, Cold Regions Science and Engineering: U.S. Army Materiel Command Cold Regions Research and Eng. Lab. Mon.

Peak, G. W., and Crook, A. G., 1967, Summary of snow survey measurements, Wyoming: Casper, Wyoming, U.S. Soil Conserv. Service, 152 p.

Peev, C. D., 1966, Geomorphic activity of snow avalanches: p. 357–368 *in* International symposium on scientific aspects of snow and ice avalanches (Davos, 1965): Internat. Assoc. Sci. Hydrology Pub. no. 69, 424 p.

Pettijohn, F. J., and Potter, P. E., 1964, Atlas and glossary of primary sedimentary structures: New York, Springer Verlag, 370 p.

Porter, S. C., and Denton, G. H., 1967, Chronology of Neoglaciation in the North America Cordillera: Am. Jour. Sci., v. 265, p. 177–210.

Potter, N., Jr., 1968, Galena Creek rock glacier, northern Absaroka Mountains, Wyoming (Abstract): Geol. Soc. America Spec. Paper 115, p. 438.

Potter, P. E., and Pettijohn, F. J., 1963, Paleocurrents and basin analysis: New York, Academic Press, Inc., 296 p.

Rapp, A., 1959, Avalanche boulder tongues in Lappland: Geog. Annaler, v. 41, p. 34–48.

—— 1960, Recent development of mountain slopes in Kärkevagge and surroundings, northern Scandinavia: Geog. Annaler, v. 42, p. 71–200.

Sigafoos, R. S., 1964, Botanical evidence of floods and flood-plain deposition: U.S. Geol. Survey Prof. Paper 485-A, 35 p.

Sigafoos, R. S., and Hendricks, E. L., 1961, Botanical evidence of the modern history of Nisqually Glacier, Washington: U.S. Geol. Survey Prof. Paper 387-A, p. A1–A20.

U.S. Forest Service, 1961, Snow avalanches, a handbook of forecasting and control measures: U.S. Dept. Agriculture Handb. no. 194, 84 p.

White, S. E., 1968, Rockfall, alluvial, and avalanche talus in the Colorado Front Range (Abstract): Geol. Soc. America Spec. Paper 115, p. 237.

Wolman, M. G., 1954, A method of sampling coarse river-bed material: Am. Geophys. Union Trans., v. 35, p. 951–956.

GEOLOGICAL SOCIETY OF AMERICA, INC.
SPECIAL PAPER 123

Radiocarbon Dating of Landslides in Southern California and Engineering Geology Implications

MARTIN L. STOUT

California State College, Los Angeles, California

ABSTRACT

Subsurface exploration of two landslides in southern California yielded carbonized wood fragments below the slide masses. One slide, just east of the active Portuguese Bend slide in Palos Verdes Hills, Los Angeles County, in rocks of the Altamira Member of the Miocene Monterey Formation, overran alluvium containing wood fragments; the radiocarbon age of the wood is 2915 ± 205 years B.P. The slide mass and scarp show minor erosion with rills and channels about 3 feet deep, and at least 25 feet of clay and silt have been deposited in the tension cracks at the head of the slide. The other slide, in southern Orange County in the Miocene-Pliocene Capistrano Formation, overran brush on the lower valley slope; the radiocarbon age of the wood is 17,180 ± 750 years B.P. Tributary valleys more than 40 feet deep have been eroded in the slide debris since movement, and more than 35 feet of sandy and silty alluvium has been deposited against the toe.

Movement of the San Juan Capistrano landslide is correlated with the maximum late-Wisconsin low stand of sea level, about 20,000 to 17,000 years B.P. postulated by Curray (1965). Greatly increased precipitation during this time, coupled with an extremely low sea level, probably accelerated the rate of downcutting in the San Juan Valley, and oversteepened valley slopes resulted in extensive landsliding. Stream erosion during this time probably established equilibrium with the lower base level.

CONTENTS

INTRODUCTION

During exploratory investigations for urban development in southern California, carbonized and sulfurized wood remains were found beneath landslides in two different areas: (1) Palos Verdes Hills, Los Angeles County, and (2) near San Juan Capistrano, Orange County (Fig. 1). Because the geologic relationships were well-known in both areas and in each the wood fragments had been overrun by slide debris, the radiocarbon ages of the carbonaceous material correspond to the maximum ages of movement for the landslides. After age determinations were made, erosion and sedimentation rates in the slide masses and surrounding area since movement could be established.

The radiocarbon ages of the two samples were determined by Geochron Laboratories, Inc. The small size of each sample yielded larger than normal errors.

ACKNOWLEDGMENTS

The writer is indebted to Moore and Taber, Engineers-Geologists, for permission to present pertinent geologic data from the two areas and to Douglas R. Brown and Perry L. Ehlig for critically reading the manuscript. Boring data were obtained from the California Division of Highways as well as from Moore and Taber, and water-well data were obtained from the California Department of Water Resources. The wood sample from Palos

Verdes Hills was collected by J. L. McNey, who also provided many photographs and a complete description of the Palos Verdes slide. The radiocarbon analytical work was supported by an institutional grant of the National Science Foundation.

DESCRIPTIVE AND ENGINEERING GEOLOGY

Palos Verdes Landslide

The Palos Verdes landslide is about 6000 feet east-southeast of the eastern margin of the famous Portuguese Bend slide in Los Angeles County (Fig. 1). The toe (or lower portion) of the slide is at approximate elevation 570 feet (173.7 m), which is apparently just below the general level of elevated marine terrace number 7 (Woodring and others, 1946). Nearby borings show that the bedrock surface below the terrace 7 deposits may have a seaward gradient no steeper than 4.5:1. Largely because of the seaward slope and the very permeable alluvium and sandy terrace deposits in the area, no ground water was found near the wood sample.

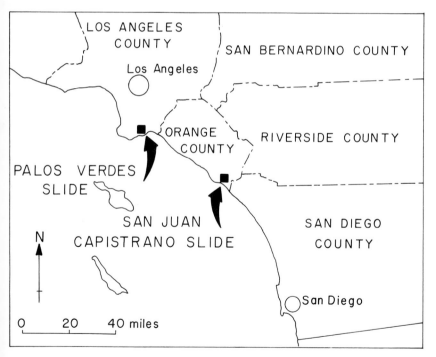

Figure 1. Location map showing Palos Verdes and San Juan Capistrano landslides, southern California.

The slide mass is underlain by diatomaceous and siliceous shales, sandstones and clayey siltstones, and bentonitic shales of the marine Altamira Member of the middle Miocene Monterey Formation (Woodring and others, 1946). The slide is essentially a composite bedding-plane failure along at least two failure surfaces that are in bentonitic beds about 9 stratigraphic feet apart. Slide movement was apparently caused principally by undercutting of southerly dipping bentonitic beds below the seaward-facing slope by a small tributary stream (Fig. 2). The slide mass covers about 6 acres and involved about 250,000 cubic yards. Several slides have occurred

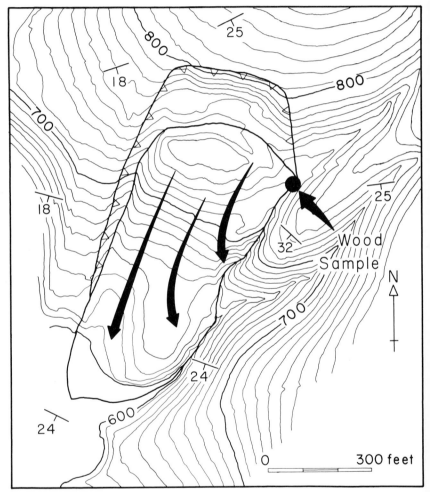

Figure 2. Geologic sketch map of Palos Verdes landslide showing location of wood sample.

within the slide debris and on the oversteepened scarp since the original movement.

Only minor erosion of the slide mass and scarp has occurred since movement; rills and channels are up to 3 feet deep. No significant amount of recent alluvium was noted near the toe or margins of the slide, although more than 25 feet of clayey material has been deposited in tension cracks at the head of the slide and slopewash from the scarp has filled all original closed depressions. For comparative purposes, a stereogram of vertical aerial photographs is included (Pl. 1, fig. 1).

The age of the wood sample is 2915 ± 205 years B.P. The wood sample was found in a bulldozer cut along the eastern margin of the main slide (Fig. 2). It was collected from a thin bed of sandy material that is believed to be alluvium of the former bed of the marginal tributary stream. The position of the wood just below the contact between slide debris and alluvium indicates that the placement of the wood probably occurred only a short time before movement of the slide, so a large difference in time of several hundred years is not considered likely.

San Juan Capistrano Landslide

The San Juan Capistrano slide is approximately 7000 feet southeast of the San Juan Capistrano mission, in the center of the town of San Juan Capistrano in the southern portion of Orange County (Figs. 1, 3). It is an area which is locally well-known for the large number of landslides; the writer has mapped about 50 slides in nearly 250 hillside acres. The slide discussed in this paper is on the eastern side of a small unnamed tributary valley which drains northward into southwesterly flowing San Juan Creek (Fig. 3).

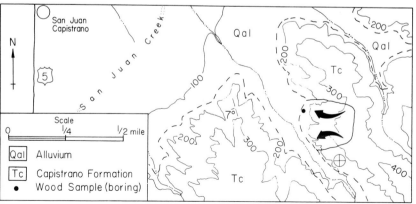

Figure 3. Geologic sketch map of San Juan Capistrano landslide. Darkened circle in slide mass is location of boring from which brush was taken.

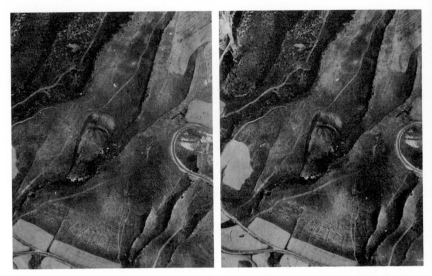

Figure 1. Stereogram showing Palos Verdes landslide: age is 2915 ± 205 years B.P. Scale of photographs is approximately 1 inch equals 1000 feet. North is to the top.

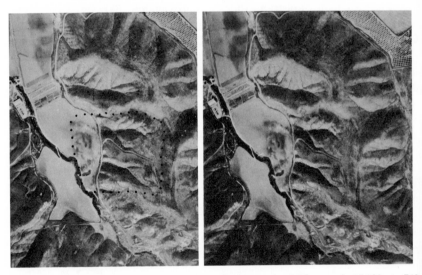

Figure 2. Stereogram showing San Juan Capistrano landslide: age is 17,180 ± 750 years B.P. Scale of photographs is approximately 1 inch equals 1300 feet. North is to the top.

STEREOGRAMS OF PALOS VERDES AND
SAN JUAN CAPISTRANO LANDSLIDES

STOUT, PLATE 1
Geological Society of America Special Paper 123

The slide is underlain by clayey siltstone and shale of the marine Miocene-Pliocene Capistrano Formation (Vedder and others, 1957). The siltstone is massive and usually weathers to shades of gray with numerous hematitic-stained joint surfaces, easily mistaken for bedding, including "pseudograded" bedding shown by varying concentrations of hematite away from, but parallel to, joints. Below 25 to 35 feet, the nonoxidized siltstone is dark gray to black; bedding can be defined by very fine grained, subaligned mica.

The slide mass covers about 24 acres and originally involved close to 2 million cubic yards. Two tributary canyons, one 41 feet deep, have been cut almost the entire length of the slide, removing more than 30,000 cubic yards of material. Although it is difficult to determine whether or not all of the erosion occurred after sliding, it is believed that most of the erosion was later because the present tributary valleys are close to the slip surface in places without apparently modifying the slip-surface configuration. Only a subtle terrace marks the original upper portion of the slide. For comparison with the Palos Verdes slide, a stereogram of vertical aerial photographs is included (Pl. 1, fig. 2).

The age of the wood is 17,180 ± 750 years B.P. The wood sample was collected in a 24-inch-diameter boring, 41 feet below natural grade (or at elevation 124 feet [37.8 m]) near the toe of the slide and about 720 feet east of the present stream course in the valley (Fig. 3). Although the wood sample was well below the ground water level in adjoining valley alluvium (ground water was at an elevation of 132 feet [40.2 m]), the boring remained dry during drilling and inspection because of the relatively impermeable clay. It is believed that the brush samples collected in the boring were originally on the valley slope (not the valley floor) and were overrun by the slide mass. This is indicated by the absence of sand and concretionary fragments so common in present-day alluvium in the area. Hence, the age of the wood represents the approximate age of slide movement.

In addition to the erosion, at least 35 feet of sandy and silty alluvium have been deposited in the valley against the toe of the slide, in essence effectively buttressing the slide mass (Fig. 4). Borings in the valley showed that the

Figure 4. Section through center of slide mass, San Juan Capistrano. Note buttressing effect of alluvium.

alluvium was not disturbed, indicating that the alluvium was deposited after slide movement. The borings also showed that very little alluvium is below the slide debris.

Significance of Carbon 14 Date for San Juan Capistrano Landslide

Curray (1965, p. 724) stated that "the maximum late-Wisconsin low stand of sea level occurred between about 20,000 and 17,000 B.P. at a level of about −120 to −125 m," based on evaluation of about 150 radiocarbon dates of shallow-water deposits and fossils. Curray's data agree well with those of Kenney (1964). The radiocarbon date of the San Juan Capistrano landslide appears to correlate well with this stillstand, indicating that widespread erosion occurred in the San Juan Valley and along its tributaries during or before this time; and, in fact, stream erosion established or nearly established equilibrium with the extremely lowered base level.

Several test borings in San Juan Valley below San Juan Capistrano establish minimum thicknesses of alluvium. Approximately 1 to 2 miles south of the San Juan Capistrano landslide, on the southeastern side of the valley, are three large landslides. One covers more than 400 acres and moved north-westerly into San Juan Valley. Borings and sections made for stability analyses in the landslide establish that alluvium extends to at least elevation –61 feet (–18.6 m). Borings were not placed directly at the toe, but the slip-surface geometry suggests that an alluvial plain existed at approximate elevation –150 feet (−45.7 m) or below at the time sliding occurred. Although this slide has not been dated, it is believed to be the same approximate age as the San Juan Capistrano slide, and all three slides appear to predate the alluvial filling in the San Juan Valley.

In addition, numerous water wells have been drilled in the alluvium of San Juan Valley. The well logs show that alluvium extends to at least elevation −100 feet (−30.5 m) 0.5 mile southwest of San Juan Capistrano (R. Blodni-kar, 1967, personal commun.). Near the present shoreline, more than 200 feet (61.0 m) of alluvium have been deposited (California Department of Water Resources, 1967, Pl. 4A). Comparing present stream gradients in the area with known elevations on the alluvial base indicates that alluvium possibly extends to −250 feet (−76.2 m) offshore from San Juan Creek. Further evidence for base level at this elevation is given by a pronounced submarine terrace off Dana Point at approximate elevation −250 feet (Emery, 1958).

Filling of San Juan Valley with alluvium produced a broad alluvial plain which is typical of other large drainage systems in southern California. Wahrhaftig and Birman (1965, p. 321) reported that postglacial sea-level rise resulted in as much as 79 to 98 feet (24 to 30 m) of alluviation in large drainage systems of the southern Coast Ranges. Poland and Piper and others (1956) and Poland and Garrett and others (1959) reported as much as 180 feet (55 m) of alluvium in the valleys of the Santa Ana, San Gabriel, and Los Angeles Rivers. About 23 miles southeast of the mouth of San Juan Creek, as much as 200 feet (61.0 m) of alluvial and estuarine deposits are below the present tidal flats at the mouth of the Santa Margarita River (California Department of Water Resources, 1967, Pl. 4A). Emery (1958) noted as much as 230 feet (70.1 m) of channel fill in the river systems crossing the Los Angeles and Ventura basins. Poland and Green (1962) reported as much as 787 feet (240 m) of alluvium in the Santa Clara Valley southeast of San Francisco, indicating that subsidence occurred as well as alluvial filling. In many other places, the depth of alluvium is not known, but its presence below sea level is suggested by broad valley floors near present sea level.

LATE WISCONSIN CLIMATE AND ITS EFFECT ON ENGINEERING GEOLOGY

It is well-established that climatic trends in southern California during the past 15,000 to 20,000 years have been toward a dryer climate at present, but unusually wet cycles may be superimposed on this general trend (B. Templeton, 1965, personal commun.). Based on a study of 1445 specimens of fruits and seeds preserved in the Rancho La Brea deposits in Los Angeles, Templeton (1964, Ph.D. dissert., Oregon State Univ.) concluded that when annual precipitation values dropped below 20 inches for extended periods, certain species either migrated to other areas or became extinct. Radiocarbon dating of a cypress tree found in the Rancho La Brea tar pits and analysis of tree rings covering a 40-year period indicates that the climate was significantly wetter 14,000 to 15,000 years B.P. (Howard, 1960; B. Templeton, 1968, personal commun.) (Table 1). The mean rainfall for this period was more than double the present mean. These local data agree well with the palynological interpretation of glacial to postglacial conditions elsewhere in the southwestern United States (Martin and Mehringer, 1965).

Several other engineering geological observations support the conclusion of an earlier, wetter climate in southern California. Thick "paleosoils" (up to 30 feet) are locally preserved as remnants in old canyons, not related to present drainage patterns, in the northern San Fernando Valley, Palos Verdes Hills, eastern Los Angeles County (Puente and San Jose Hills), and in the Laguna Beach area where active present-day erosion has not removed them. Other thick "paleosoil" remnants are being found in southern California in large artificial cuts. These suggest a more rapid development of soil in the recent past than at present, probably as a result of increased precipitation, because other factors of soil formation have remained essentially constant. Soil and rock creep in western Los Angeles County have also been more active in the recent past than at present (Stout, 1965). Again other local factors

TABLE 1.   COMPARATIVE PRECIPITATION VALUES IN LOS ANGELES, 1877 TO 1966 AND APPROXIMATELY 15,000 YEARS B.P.*

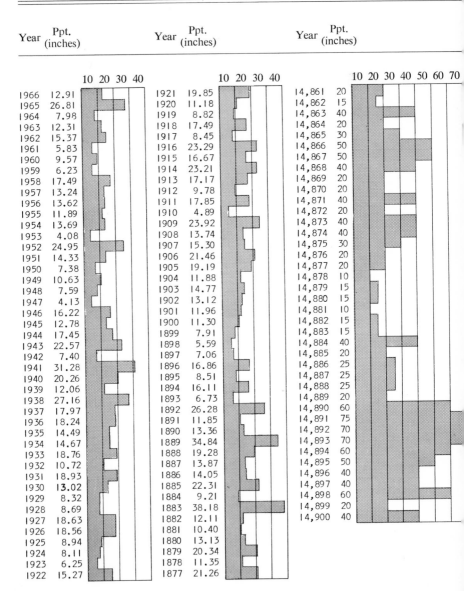

Year	Ppt. (inches)	Year	Ppt. (inches)	Year	Ppt. (inches)
1966	12.91	1921	19.85	14,861	20
1965	26.81	1920	11.18	14,862	15
1964	7.98	1919	8.82	14,863	40
1963	12.31	1918	17.49	14,864	20
1962	15.37	1917	8.45	14,865	30
1961	5.83	1916	23.29	14,866	50
1960	9.57	1915	16.67	14,867	50
1959	6.23	1914	23.21	14,868	40
1958	17.49	1913	17.17	14,869	20
1957	13.24	1912	9.78	14,870	20
1956	13.62	1911	17.85	14,871	40
1955	11.89	1910	4.89	14,872	20
1954	13.69	1909	23.92	14,873	40
1953	4.08	1908	13.74	14,874	40
1952	24.95	1907	15.30	14,875	30
1951	14.33	1906	21.46	14,876	20
1950	7.38	1905	19.19	14,877	20
1949	10.63	1904	11.88	14,878	10
1948	7.59	1903	14.77	14,879	15
1947	4.13	1902	13.12	14,880	15
1946	16.22	1901	11.96	14,881	10
1945	12.78	1900	11.30	14,882	15
1944	17.45	1899	7.91	14,883	15
1943	22.57	1898	5.59	14,884	40
1942	7.40	1897	7.06	14,885	20
1941	31.28	1896	16.86	14,886	25
1940	20.26	1895	8.51	14,887	25
1939	12.06	1894	16.11	14,888	25
1938	27.16	1893	6.73	14,889	20
1937	17.97	1892	26.28	14,890	60
1936	18.24	1891	11.85	14,891	75
1935	14.49	1890	13.36	14,892	70
1934	14.67	1889	34.84	14,893	70
1933	18.76	1888	19.28	14,894	60
1932	10.72	1887	13.87	14,895	50
1931	18.93	1886	14.05	14,896	40
1930	13.02	1885	22.31	14,897	40
1929	8.32	1884	9.21	14,898	60
1928	8.69	1883	38.18	14,899	20
1927	18.63	1882	12.11	14,900	40
1926	18.56	1881	10.40		
1925	8.94	1880	13.13		
1924	8.11	1879	20.34		
1923	6.25	1878	11.35		
1922	15.27	1877	21.26		

* Data are based on U.S. Weather Bureau records, and analysis of tree-ring data (Temple from cypress tree in La Brea tar pits.

remain constant, so apparently increased precipitation has been the most influential factor. Depths of soil and rock creep in the greater Los Angeles and southern California areas are as great as 12 to 30 feet. Soil or rock creep, or both, is common on most natural hillside areas in those portions of Los Angeles and Orange Counties underlain by sedimentary rock.

SIGNIFICANCE OF RADIOCARBON DATING OF LANDSLIDES FOR ENGINEERING GEOLOGISTS

This study shows that terminology related to times of movement of landslides, if used at all, must be carefully spelled out in engineering geology reports. For example, to many geologists, both of the landslides discussed in this paper would be categorized as "recent"; to others, they would be "ancient" because they occurred before the last decade or century. It seems much more desirable to give an estimate of the approximate time of movement based on erosional characteristics and the like, if no other information is available, than to express the age of movement in nonquantitative terms. Even rounding off to the nearest 100 or 1000 years is better than ancient, medieval, young, old, recent, and so forth. Even historic and prehistoric have their share of problems because of inadequate data regarding the cutoff point in southern California (Moriarity, 1967).

Erosional characteristics of landslides appear to be useful criteria for correlation of ages in similar climatic areas. Although the type of bedrock is important, its effect is apparently lessened after failure. An older landslide in Palos Verdes Hills, just east of the slide discussed earlier, is also underlain by the Altamira Member of the Monterey Formation. The slide has recently been dated at 16,200 ± 240 years B.P. (P. Ehlig, 1968, personal commun.). Despite different types of bedrock, the erosional characteristics of this slide are similar to the San Juan Capistrano slide except for a more pronounced terracing in the upper portion of the Palos Verdes slide. With continued erosion of these slides, it appears that the topographic expression of the landslides might be lost in another few thousand years. Thus, most landslides recognized as topographic features in similar climatic areas have probably formed within the time range of carbon 14 (up to 50,000 years B.P.).

After the age of the slide is established, the geologic history after movement can be critical in the final engineering and stability analyses: has alluvium been deposited at the toe as a natural buttress; has slide debris been eroded from the toe, decreasing the stability of the slide; has erosion removed portions of the upper slide mass, increasing the stability?

Not all landslides are as favorably treated by time as the San Juan Capistrano slide where natural buttressing of the slide mass has occurred, as well as a sizable reduction in driving force because of the large quantity of material removed from the slide mass by erosion. Calculations using strength values of 200 psf cohesion and an angle of internal friction of 8°, based only on the buttressing effect of alluvium, give an increase of 0.35 in the factor of safety. Calculations based on a section through the deepest level of erosion give an increase in the factor of safety of 0.46. If both buttressing by alluvium and maximum removal of material by erosion are considered, the factor of safety is increased by approximately 0.64. This latter figure would be true for only that portion of the slide just below the canyons that has been eroded since slide movement; the over-all factor of safety for the entire slide is probably about 1.40.

It is important that the engineering geologist working with older landslides along present coast lines or older large landslides along alluviated valley floors recognize the possibility that failure could have occurred when sea level was much lower. In these places, the slip surface would *not* necessarily toe out at present sea level or at the present surface of the alluviated valley. Slosson and Cilweck (1966) described a landslide on Santa Catalina Island with a slip surface apparently well below sea level, and, based on sea-level fluctuations in the recent past, they concluded that the slide movement may have started about 17,000 years ago. Substantial buttressing by alluvial or marine deposits or erosion by surficial processes in the toe area of the slide now covered by water can easily modify the factor of safety, and these features must be considered in the stability analysis of the entire slide.

REFERENCES CITED

California Department of Water Resources, 1967, Ground water occurrence and quality: San Diego Region: California Dept. Water Resources Bull. 106-2, 235 p.

Curray, J. R., 1965, Late Quaternary history, continental shelves of the United States, p. 723–735 *in* Wright, H. E., Jr., and Frey, D. G., *Editors*, The Quaternary of the United States: Princeton, New Jersey, Princeton Univ. Press, 922 p.

Emery, K. O., 1958, Shallow submerged marine terraces of southern California: Geol. Soc. America Bull., v. 69, p. 39–60.

Howard, H., 1960, Significance of carbon-14 dates for Rancho La Brea: Science, v. 131, p. 712–714.

Kenney, T. C., 1964, Sea level movements and the geologic histories of the postglacial marine soils at Boston; Nicolet, Ottawa, and Oslo: Géotechnique, v. XIV, p. 203–230.

Martin, P. S., and Mehringer, P. J., Jr., 1965, Pleistocene pollen analysis and biogeography of the southwest, p. 433–451 *in* Wright, H. E., Jr., and Frey, D. G., *Editors*, The Quaternary of the United States: Princeton, New Jersey, Princeton Univ. Press, 922 p.

Moriarity, J. R., 1967, Transitional pre-desert phase in San Diego County, California: Science, v. 155, p. 553–556.

Poland, J. F., Garrett, A. A., and Sinnott, A., 1959, Geology, hydrology, and chemical character of ground waters in the Torrance–Santa Monica area, California: U.S. Geol. Survey Water-Supply Paper 1461, 425 p.

Poland, J. F., and Green, J. H., 1962, Subsidence in the Santa Clara Valley, California, a progress report: U.S. Geol. Survey Water-Supply Paper 1619-C, 16 p.

Poland, J. F., and Piper, A. M., and others, 1956, Ground-water geology of the Coastal Zone, Long Beach–Santa Ana area, California: U.S. Geol. Survey Water-Supply Paper 1109, 162 p.

Slosson, J. E., and Cilweck, B. A., 1966, Parson's Landing landslide, a case history including the effects of eustatic sea level changes on stability: Eng. Geology (AEG), v. 3, p. 1–9.

Stout, M. L., 1965, Gravity folds in the Modelo Formation, western Los Angeles County, California: Geol. Soc. America Bull., v. 76, p. 967–970.

Vedder, J. G., Yerkes, R. F., and Schoellhamer, J. E., 1957, Geologic map of the San Joaquin Hills–San Juan Capistrano area, Orange County, California: U.S. Geol. Survey Oil and Gas Inv. Map OM 193.

Wahrhaftig, C., and Birman, J. H., 1965, The Quaternary of the Pacific Mountain System in California, p. 299–340 *in* Wright, H. E., Jr., and Frey, D. G., *Editors*, The Quaternary of the United States: Princeton, New Jersey, Princeton Univ. Press, 922 p.

Woodring, W. P., Bramlette, M. N., and Kew, W. S. W., 1946, Geology and paleontology of Palos Verdes Hills, California: U.S. Geol. Survey Prof. Paper 207, 145 p.

GEOLOGICAL SOCIETY OF AMERICA, INC.
SPECIAL PAPER 123

Glacial Geology of the Lower Alatna Valley, Brooks Range, Alaska

THOMAS D. HAMILTON
University of Alaska, College, Alaska

ABSTRACT

Glaciers originating in the central Brooks Range extended south into the lower Alatna Valley during three major episodes of Illinoian and Wisconsin glaciation. During the oldest glaciation, coalescing glaciers formed a piedmont ice sheet that extended 60 miles south of the range and covered most of the Koyukuk Lowlands. The later Kobuk Glaciation, subdivided into two stades, was marked by a smaller lobe that terminated in the Koyukuk Lowlands, receded, then readvanced to a frontal position within the Alatna Valley 25 miles closer to the Brooks Range. Drift of the Itkillik Glaciation, which marks the final period of ice advance into the lower Alatna, is subdivided into four moraine belts. Siruk Creek moraines, the oldest and most extensive Itkillik deposits, form an arcuate belt that extends 30 miles south of the range front. The later Chebanika moraine marks a brief readvance or stillstand of glacier ice farther upvalley. Ice then stagnated in the Helpmejack Lakes area, and massive bodies of ice-contact stratified drift were deposited in a broad belt 8 to 15 miles south of the Brooks Range. Moraines at the range front represent the last major event of the Itkillik Glaciation.

Subdivisions and correlations of the glacial sequence are based on extent and position of the drift sheets, inferred glacier regimens, postglacial modification of glacial deposits, radiometric dates, and ice limits within the upper Alatna Valley. The four episodes of Itkillik glaciation within the Alatna drainage system appear to be equivalent to four Itkillik stades in the Anaktuvuk Valley, and radiocarbon dates from both areas indicate a late Wisconsin age for Itkillik deposits. The preceding ice advance, correlated with the Kobuk Glaciation farther west, may represent the early Wisconsin. The oldest glaciation is correlated with an ice advance of Illinoian age that filled the Kobuk Valley and extended west into Kotzebue Sound.

CONTENTS

INTRODUCTION

Although multiple episodes of Pleistocene glaciation have been reported from several parts of northern Alaska, most regional correlations and age estimates of the glacial sequence have been speculative. Few radiometric dates are available for glacial events of the region, and, where such dates are available, usually only a short segment of late Pleistocene history is represented. Detailed studies of glacial events and related environmental changes have been carried out primarily in parts of the northern Brooks Range (Detterman and others, 1958; Porter, 1964, 1967; Holmes and Lewis, 1965) and in the Kotzebue Sound–Seward Peninsula area (Hopkins, 1963, 1967; McCulloch, 1967; McCulloch and Hopkins, 1966; Sainsbury, 1967). Regional application of the data obtained in these studies has been handicapped by lack of detailed investigations in intervening areas, by the discontinuous nature of the alpine drift sheets, and by the extreme environmental diversity of the region.

The present study was initiated with the objectives of defining the glacial history of a valley system in the southern Brooks Range, dating the glacial events as fully as possible, and discovering means by which these events could be correlated directly with the partial chronologies available for regions farther to the north and to the west. The Alatna Valley was chosen for detailed study because of its proximity to the Killik and Nigu drainage systems, which extend north to the Arctic flank of the Brooks Range, and to the Kobuk Valley, which trends west to Kotzebue Sound.

Field studies in the Alatna Valley were started in 1962 and continued during 1963 and 1965. Three major glaciations provisionally correlated with

Illinoian, early Wisconsin, and late Wisconsin time are represented by discrete drift sheets beyond the southern range front. The two most recent glaciations were marked by multiple stades. Physical continuities were found to exist at the valley head between former glaciers in the Alatna system and glaciers that extended north to the Arctic Foothills of the Brooks Range. Drift of the oldest glaciation recognized in the Koyukuk Lowlands south of the Alatna Valley can be traced westward into the oldest known drift sheet of the Kobuk Valley–Kotzebue Sound region. Radiocarbon dating of eight organic samples recovered from the Alatna Valley and the Koyukuk Valley provides additional data on glacial and postglacial events.

ACKNOWLEDGMENTS

Investigations of the Alatna Valley were completed as part of a doctoral program at the University of Washington under the guidance of Professor Stephen C. Porter. Field research was supported by grants from the Geological Society of America, the National Science Foundation, and the Arctic Institute of North America under contract from the Office of Naval Research. Radiocarbon dates to supplement those obtained from a commercial laboratory were provided by A. W. Fairhall, University of Washington, and by Meyer Rubin, U.S. Geological Survey.

Continuing studies in the Kobuk and Killik Valleys have been supported by financial grants from the National Science Foundation and facilities provided through Brown University and the University of Alaska.

I am greatly indebted to Stephen C. Porter, Robert F. Black, David M. Hopkins, and Troy L. Péwé for council during the course of the study and for critical reviews of the present paper.

REGIONAL SETTING

From a low divide close to the northern flank of the Brooks Range, the Alatna River flows southward through the mountain belt and its southern foothills into the broad lowlands of the Koyukuk Valley (Fig. 1). The upper course of the Alatna, which extends 85 miles to the southern range front, occupies a deep glacial trough. The lower Alatna Valley, extending 55 miles farther south through foothills and lowlands beyond the Brooks Range, is characterized by gentler relief, lower gradient, and relatively unconfined drainage courses subject to repeated glacial displacements. Valley glaciers flowing southward through the Alatna and neighboring drainage systems tended to spread laterally into individual or coalescent piedmont lobes beyond the southern flank of the Brooks Range; glacier deposition was dominant over erosion throughout this region (Pl. 1).

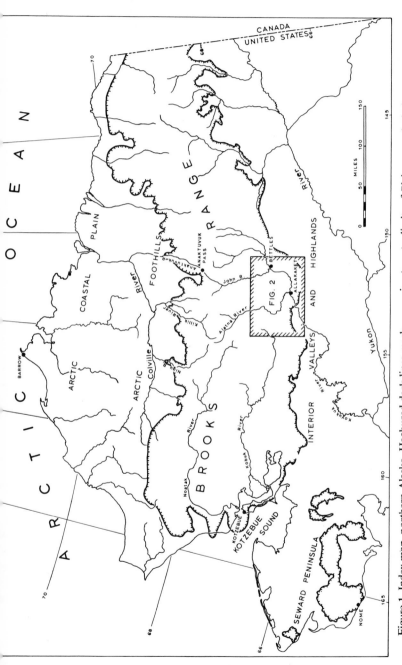

Figure 1. Index map, northern Alaska. Hachured dark lines mark approximate outer limits of Pleistocene glaciation (*after* Coulter and others, 1965).

OLDEST GLACIATION

The earliest known glacial advance is marked by truncated mountain spurs at high altitudes in the south-central Brooks Range and by ice-scoured uplands, glacial-drainage changes, and possible drift remnants farther south. Ice filled the Alatna Valley and the upper Koyukuk Valley, coalescing to form a lobe about 40 miles wide that extended 60 miles or more south of the Brooks Range (Pl. 1). Extent and distribution of the ice limits suggest correlation with glacial deposits of Illinoian age in the Kotzebue Sound area.

Evidence for this glacial event was recognized first by Schrader (1904, p. 90), who found rounded and probably ice-scoured uplands to altitudes of about 1800 feet through much of the Koyukuk Valley north of Kanuti River. Highly weathered erratic boulders and a possible moraine in the upper Kanuti drainage later were described by Eakin (1916, p. 54). A probable drift border separating hummocky ground moraine to the north from relatively smooth hillslopes to the south can be traced on aerial photographs for more than 100 miles along the southern margin of the Koyukuk Lowlands and westward into the Kobuk Valley (Pl. 1). This drift border approximately coincides with the inferred limit of "glacial advances of middle to early Pleistocene age" shown on a recent glacial map of Alaska (Coulter and others, 1965).

Altitudinal limits of glacier activity suggest that the ice surface rose northward from an altitude of about 1000 feet in the terminal zone to roughly 2000 feet near the mouth of the Alatna Valley and 4000 feet at the southern margin of the Brooks Range (Pl. 2). Irregular hummocky terrain and associated morainelike ridges extend across Kanuti Flats, and uplands between Allakaket and Kanuti Canyon appear to have been abraded by glacier ice to altitudes of at least 1000 feet. Ridges that rise to nearly 2000 feet along both flanks of the Alanta Valley near its mouth appear to have been overridden by glacier ice. Their smooth crests are interrupted by broad transverse channels where divides probably were crossed by ice and melt water, and some appear streamlined in a southeastward direction by glacial abrasion. Farther east, glaciers of comparable age appear to have overridden divide areas as high as 2000 feet in altitude through the foothill belt between the Alatna and John Rivers. Northward along the Alatna Valley, ice-scoured uplands rise to higher altitudes, and mountain spurs near the southern flank of the Brooks Range have been truncated to an upper level of about 4000 feet (Pl. 2). The limit of glacial abrasion here lies more than a thousand feet higher than the upper limits of later glaciations. It declines in altitude southward and, in both height and inclination, appears continuous with the oldest ice limit of the Koyukuk Lowlands.

Glacier ice occupied the Koyukuk Valley at least as far west as Mentanontli River, extending across the Kanuti Flats and probably diverting the lower portion of Kanuti River from a preglacial course to an ice-marginal position through Kanuti Canyon (Pl. 1).

KOBUK GLACIATION

The Kobuk Glaciation, named by Fernald (1964) after extensive drift exposures along the Kobuk River, is the oldest glacial event for which direct evidence is available within the lowlands of the Alatna drainage system. Ground moraine correlated with this glaciation is widespread through the lower Alatna Valley and, where cut by the Alatna River and adjacent parts of the Koyukuk, steep river bluffs expose till, outwash, and ice-contact stratified drift. End moraines in the Koyukuk Valley and lateral moraines within the lower Alatna near its mouth were deposited during the maximum ice advance of the Kobuk Glaciation. During a subsequent recessional stade, moraines and outwash were deposited farther north in the Alatna Valley (Pl. 3). Radiocarbon analyses of organic samples taken from drift of Kobuk age in the Alatna and Kobuk Valleys yield infinite dates; the drift therefore is older than the time span of the late Wisconsin. It may be of early Wisconsin age, or, alternatively, it may represent late episodes of the Illinoian glaciation.

Maximum Advance

River bluffs exposing drift of Kobuk age form a continuous scarp more than 100 feet high that extends along the northern side of the Koyukuk River from near the Alatna confluence downstream to the vicinity of Bergman Creek (Pl. 3). Comparable bluffs along the southern side of the Koyukuk River extend about 15 miles farther upvalley. The bluffs yield useful exposures only where recent river cutting has occurred; elsewhere their faces are obscured by slumps, mudflows, and vegetation. Other evidence for the maximum advance of the Kobuk Glaciation includes (1) drainage diversions and deposits of proglacial lakes within the Koyukuk Lowlands, (2) moraines, kame terraces, and ice-marginal drainage channels within the lower Alatna Valley, and (3) truncated mountain spurs at intermediate altitudes within the central Brooks Range.

Ice Limits. End moraines outline a glacial lobe that extended about 15 miles beyond the mouth of the Alatna Valley and terminated in the northern Kanuti Flats (Pl. 3). To the west, the glacier probably was in contact with glacier ice that extended south from headward parts of the Kobuk Valley. To the east, the glacier was confined by uplands bordering the Alatna Valley and was separate from glacier lobes in the upper Koyukuk drainage (Pl. 1). Moraine ridges north of the Koyukuk between the Alatna Valley and Henshaw Creek trend eastward along the surface of a drift belt nearly 3 miles wide and more than 200 feet above the adjacent Koyukuk flood plain. Drift here forms a pronounced embankment against the base of Double Point Mountain, then curves southeast to intersect the Koyukuk River about 16 miles east of Allakaket (Pl. 3). Moraines west of the Alatna's mouth intersect the Koyukuk near Bergman Creek. This moraine belt displays generally irregular surface relief near its outer margin, but closer to the Alatna it is marked by a series of elongate ridges and parallel drainage lines. Three concentric moraine ridges cross the northern Kanuti Flats and extend northward to join the moraines that enter the Koyukuk from both flanks of the Alatna Valley.

Undifferentiated drift, which probably includes both lateral moraines and kame terraces, extends along the base of the northeastern flank of the lower Alatna Valley. The drift stands at altitudes of about 600 to 900 feet near the mouth of the Alatna, forming elongate ridges and benches that produce a belt of flattened slopes near the base of the valley side. Farther north, near Sinyalak Creek, drift and ice-marginal drainage channels are associated with ridge spurs truncated to about 1600 feet altitude by glacial abrasion (Pl. 2). Streamlined bedrock knobs and molded drift bodies occur in the area between Norutak Lake and the Alatna River (Pls. 1, 3). Much of the drift forms medial moraine deposited between glaciers in the Alatna and Kobuk Valleys.

Truncated mountain spurs at intermediate altitudes across the central Brooks Range are believed to represent glacial erosion during Kobuk time (Pl. 2). Crests of these glacial facets attain a maximum altitude of about 5000 feet near the head of the Alatna Valley and consistently decrease in altitude north and south from this probable former ice divide. The reconstructed glacier profile intersects the southern flank of the Brooks Range at an altitude of about 2500 feet. Its projection farther south intersects crests of glacial facets along both sides of the lower Alatna Valley and appears to coincide with the maximum advance of the Kobuk Glacier into the Koyukuk Lowlands.

Stratigraphy. Inferred stratigraphic relations of glacial and postglacial sediments exposed in bluffs along the Koyukuk River are shown in Figure 2. The reconstruction is based on four sections (Pl. 3; Table 1) measured in areas where recent river cutting has exposed the entire thickness of the bluff deposits. Sparse exposures elsewhere along the bluffs permit less detailed study, but occasionally they allow general observations of thickness and character of the major stratigraphic units.

A conspicuous bluff opposite Allakaket (locality 1) exhibits 70 feet of till and outwash overlain by 55 feet of eolian, lacustrine, and organic sediments. Compacted clayey basal till is overlain by sandy till that shows little compaction and contains a higher proportion of waterworn sediment. Stratified deposits between the two tills consist of well-sorted sand that coarsens upward and is succeeded by outwash gravel. Deposits overlying the upper till consist of eolian silt and fine sand, organic-rich lacustrine silt, and stratified peat.

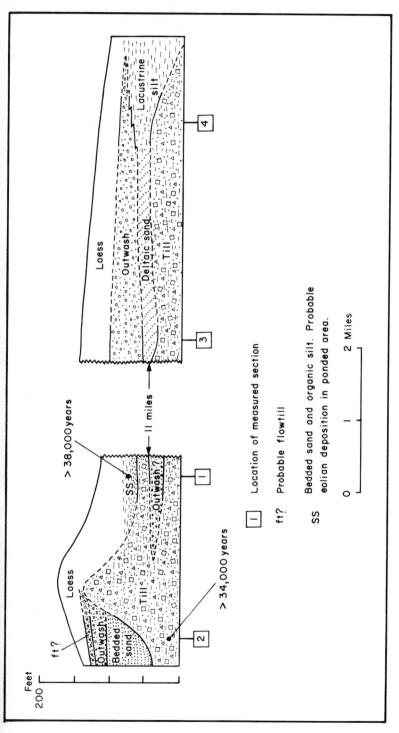

Figure 2. Inferred relations of Kobuk-age drift (maximum advance) exposed in bluffs along the Koyukuk River. *See* Figure 4 for locations of measured sections.

TABLE 1. MEASURED SECTIONS, KOYUKUK BLUFF EXPOSURES

Locality 1. Northern side of Koyukuk River opposite Allakaket

Unit	Thickness (feet)	Lithology
A	2	Peat and silt; interstratified. Penetrated by roots of living vegetation.
B	25	Brown eolian silt; nonstratified. Contains plant remains and gastropod shells.
C	10	Gray organic silt. Probable eolian deposition in pond environment.
D	2	Peat and fine eolian sand; interstratified. Yellowish oxidized sand overlain by thinner peat and sand horizons. Older than 38,000 years (UW-85).
E	10	Gray fine sand; extremely well sorted, nonstratified.
F	15	Light-gray till; oxidized on surface. Stones to boulder size in sandy matrix.
G	5	Gravel; coarse sand to cobble size; poorly sorted; subrounded.
H	20+	Gray till; stones to boulder size in clayey matrix. Indurated and slabby, becoming blocky where exposed to weathering. Slabby partings have brown oxide coatings. Contains sand horizons of probable fluvial origin. Base is covered by lag deposits along river bank.

Locality 2. Northern side of Koyukuk River 4 miles southwest of Allakaket

Unit	Thickness (feet)	Lithology
A	15	Silt; eolian.
B	12	Flowtill(?); stones to boulder size in clayey-silt matrix. Little compaction, and contact with underlying unit shows no mixing or bedding disturbance.
C	31	Gravel; sand to cobble size.
D	59	Bedded sand; very well sorted. Contains laminated silt near base. Bedding shows slump structures characteristic of ice-contact deposits.
E	50+	Gray till; sparse stones in clayey-silt matrix. Indurated. Slickensided shear planes dip toward river edge. Contains wood fragments near base that are older than 34,000 years (GX0524). Base covered by modern lag deposits along river bank.

Locality 3. Southern side of Koyukuk River 12 miles east of Allakaket

Unit	Thickness (feet)	Lithology
A	45	Silt; dominantly eolian. Gleyed, and exhibits some horizontal stratification. Contains organic horizons.
B	50	Sand and gravel; outwash.
C	25	Fine sand; cross-bedded, with beds dipping generally south to southeast. Probably deltaic.
D	25+	Gray till; stones to cobble size in clayey matrix. Upper contact rises toward west.

TABLE 1 (Continued)

Locality 4. Southern side of Koyukuk River 15 miles east of Allakaket

Unit	Thickness (feet)	Lithology
A	30	Silt; grayish and mottled at base, somewhat oxidized above.
B	20	Outwash gravel; pebbles dominant, but many sand lenses present. Bedding dips east to southeast.
C	30	Fine sand; very well sorted. Grades eastward into fine sand with clayey-silt interbeds then into highly organic lacustrine silt.
D	45+	Dark-gray till. Upper part is nonsorted, contains thin sand layers, and consists of small stones in clayey-silt matrix. Lower part contains stones to boulder size, exhibits some bedding, and has been washed free of all fines below coarse-sand size.

Locality 2, along the northern bank of the Koyukuk River 4 miles southwest of Allakaket, contains till, outwash, and ice-contact stratified drift overlain by loess. The basal till is compact and has a clayey matrix. A thin upper till-like deposit (Table 1, unit B) is more silty, is less compact, and overlies undisturbed outwash gravel; it probably represents flowtill (Hartshorn, 1958) rather than the product of a glacier readvance. Bedded sand beneath the outwash contains slump structures characteristic of stratified ice-contact sediments. Loess, which caps the section, thins southwestward, where it may have been subjected to erosion and mass wasting along the margin of Bergman Creek.

Bluffs upriver from Allakaket expose till and outwash that thin eastward and pass into thick lacustrine deposits. Locality 3 (Pl. 4) exhibits clayey till overlain by a thick sand unit, probably deltaic, in which beds dip south to southeast. Outwash and loess overlie the sand. Locality 4 contains a basal till that lacks particles smaller than coarse sand. This till is overlain by well-sorted fine sand that may represent eolian deposition in a lacustrine environment. Outwash above the sand is both thinner and finer in texture than at locality 3; bedding dips southeast, indicating a source area in the direction of the lower Alatna Valley. Exposures farther east are largely obscured by slumps and mudflows but show the same general stratigraphic sequence and eastward facies trends. The basal till here is unable to support steep erosional scarps, possibly because it was less compacted and because more fines were washed out. This till thins eastward and probably terminated close to the present margin of the bluffs (Pl. 3). Overlying lacustrine deposits thicken eastward and may interfinger with the till. Decrease in grain size from fine sand to silt occurs beyond the eastern limit of deltaic fore-set bedding. Overlying outwash thins and

LOCALITY 3, KOYUKUK RIVER BLUFFS

Till (T) forms conspicuous flows; lacustrine and deltaic sand (S) supports nearly vertical faces; outwash (OW) is marked by pronounced bedding. Loess and muck (LM) cap the section. Exposure stands 145 feet high.

becomes finer in texture eastward and appears to interfinger with the lacustrine deposit. Eolian silt, which caps the section, also thins eastward from a maximum of 45 feet at locality 3 to less than 30 feet east of locality 4.

Sediments exposed in the Koyukuk bluffs show the following general relations in the successive major units.

(1) The basal till thins toward the eastern and western margins of the bluffs and was modified by melt-water streams or by contact with bodies of standing water in these areas.

(2) Lacustrine deposits overlie the basal till near the margins of the bluffs; their deposition partly coincided with continued glacier activity in the central part of the area. The western deposit (locality 2) may be of limited extent; the eastern deposit probably represents the margin of a large lake dammed by drift and glacier ice.

(3) Dips of bedding planes toward the west at localities 1 and 2 and eastward at localities 3 and 4 indicate transport of outwash both up and down the Koyukuk Valley from a central location near the Alatna confluence.

(4) Eolian silt is interbedded with peaty layers representing periods of little wind deposition. Laminated organic deposits also are present, indicating probable thaw-lake development in the silts. The peat horizons within the eolian sequence imply that a long period of sporadic wind deposition followed ice recession from the area.

Drainage Features. Two large lakes were dammed by advance of the Kobuk glacier into the Koyukuk Lowlands (Pl. 1). The most extensive lake, which stood at an altitude of about 600 feet, covered more than 300 square miles of the lower South Fork and adjacent parts of the Koyukuk Valley. Its former position, indicated by lacustrine sediments in the bluff exposures, coincides with adjacent surface areas of low relief and poor drainage. The lake was formed when the Koyukuk River and its tributaries were dammed by glacier ice that extended southeast from the mouth of the Alatna Valley; it was fed by streams from parts of the Koyukuk drainage system and probably received melt water and sediments both from the Alatna Valley glacier and from a piedmont glacier that extended down the Koyukuk from its upper tributaries (Pl. 1). The lake probably drained south into Kanuti Flats along the abandoned channel shown on Plate 3. Normal drainage down the Koyukuk was resumed following glacial recession.

A second glacial lake, covering about 200 square miles in the southern Kanuti Flats, formed when the Alatna Valley glacier dammed probable northward drainage of the Kanuti River and its tributaries. This lake had a surface altitude of about 550 feet, probably received overflow drainage from the larger glacial lake to the north, and drained west through Kanuti Canyon. Deepening of this outlet probably lowered the lake surface prior to glacier recession from the Koyukuk Lowlands, and Kanuti River never resumed its former course north into the Koyukuk.

Channels trending west and southwest at 700- and 800-foot altitudes cross divides between Allakaket and Kanuti Canyon (Pl. 3) and probably are relics of glacial drainage toward Mentanontli River and downvalley parts of the Koyukuk. Within the lower Alatna Valley, northward-trending tributaries to Siruk Creek are separated from the heads of Discovery, Henry, and Bergman Creeks by a series of low lake-filled passes that may represent segments of the terminal moraine (Pl. 3). Much of the Siruk Creek drainage prior to glacial diversion probably flowed southeast to join the Koyukuk.

Recessional Stade

Following recession from the Koyukuk Lowlands, the Alatna Valley glacier halted at or readvanced to a new terminal position near Sinyalak Creek (Pl. 3). The modern river here cuts into an extensive moraine embankment just upvalley from Sinyalak Creek (Pl. 3, locality 6), and prominent bluffs farther downstream exhibit till, outwash, and lacustrine sediments in a series of exposures that appear to cross the outermost moraine of the recessional stade (Pl. 3, locality 5). Morainal topography is most distinctive near the valley center. Farther to the southwest and northeast, it tends to merge with deposits of the preceding ice advance.

Moraines. The outermost moraines of the recessional stade form an embankment against bedrock ridges west of the Alatna River downvalley from Sinyalak Creek (Pl. 3). Three broad morainal ridges diverge from this embankment near the southwestern side of the Alatna and curve northeast to intersect the river. A similar drift embankment stands against the southern flank of the bedrock ridge between the Alatna River and Sinyalak Creek. Its upper surface slopes east from an altitude of about 1100 feet near the northwestern end to slightly less than 800 feet where the eastern ridge tip has been truncated by Sinyalak Creek.

Moraines, kame terraces, and irregular bodies of ice-contact stratified drift flank both sides of Sinyalak Creek at altitudes of about 800 to 1000 feet and extend southeast from near the creek's mouth as a broad ridge about 60 feet high. Both trend and altitude indicate a probable connection with the outermost moraines southwest of the Alatna River (Pl. 3), but stream erosion has removed all drift from the intervening area.

Stratigraphy. Till, outwash, and lacustrine sediments are exposed in bluffs about 100 feet high at locality 5. The till comprises most of the upvalley end of the section and extends beneath outwash farther to the southeast. Downvalley, a small body of lacustrine sediment is confined between the drift and the northern flank of a bedrock ridge. Five feet or more of interstratified loess and peat cap the section. Inferred relations of the major stratigraphic units are shown in Figure 3. Extensive exposures are not present along the bluff face, and the presence of permafrost at shallow depths usually prohibits deep excavation. Approximate thickness of the units and the locations of their contacts were inferred from small, infrequent exposures and by the use of indirect lines of evidence as outlined in Table 2.

A section cut into the upper part of the bluff near its downvalley end (Fig. 3, section 5) exposed 12 feet of stratified peat and clayey silt above laminated lacustrine sediments (Table 3). The lowest unit exhibits apparent rhythmic deposition of paired couplets consisting of alternating organic and inorganic layers. Stratified peat and clayey silt above the lacustrine sediments probably record a marshy reducing environment during accumulation of peat and loess. A former permafrost table about 7 feet deep is indicated by the contrast between fresh unflattened wood fragments below this level and flattened decayed wood at shallower depths. Massive accumulation of iron oxide occurs directly above the inferred former permafrost table. The modern permafrost table lies within 1 to 2 feet of the ground surface.

Farther upvalley, more extensive till exposures are present along three south-facing cutbanks of the Alatna River (locality 6). Moraine embankments beyond these cutbanks rise to about 1000 feet altitude against the flank of a bedrock ridge that separates the Alatna from Sinyalak Creek. The cutbanks expose as much as 40 feet of gray stony till above bedrock that crops out along the river edge. Angular fragments of the bedrock have been displaced down-valley by glacier ice to form a basal lodgment till with stones oriented parallel to the original bedding. Higher parts of the till contain subangular stones to

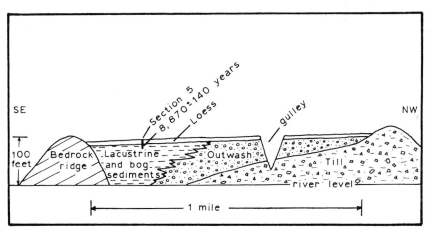

Figure 3. Inferred relations of Kobuk-age drift (recessional stade) at locality 5 near Sinyalak Creek.

TABLE 2. GEOMORPHIC INDICATORS OF PRINCIPAL STRATIGRAPHIC UNITS, SINYALAK CREEK
AREA, LOWER ALATNA VALLEY

Till	Slump blocks, mudflows, and dense thickets of willows and alders cover most of bluff face. Upper surface is irregular, poorly drained, and underlain by permafrost at shallow depth. Bluff face is convex in profile, meeting upper surface in smooth curve.
Outwash	Steep brushy flanks with large slump blocks. Flat upper surface, meeting bluff face with sharp angular junction. River bank at till-outwash contact exhibits abrupt termination of mudflows and pronounced downvalley increase in sand-sized particles relative to silt and clay. Lag concentrates show decrease in subangular tillstones and a corresponding increase in rounded particles downvalley across the till limit.
Lacustrine sediments	Upper surface level and very poorly drained, with permafrost at shallow depth. Thaw lakes are present. Bluff face subject to very extensive slumping and bears dense plant cover. River edge at lacustrine-outwash contact shows downvalley decrease in sand-sized particles relative to fines. Lag concentrate of scattered stones contrasts with more continuous lag pavement upvalley across the outwash contact.
Bedrock	Angular, frost-shattered detritus of local lithology mantles bluff face, underlies vegetation mat which covers upper surface, and forms lag concentrate along river edge below height of spring flood stage. Shallow riffles extend out into river, and bedrock is visible beneath water surface here.

cobble size in sandy and clayey lenses; lag concentrates along the river edge consist of subrounded pebbles and cobbles with a few small boulders. Most stones are graywackes and conglomerates, rock types that crop out south of the Brooks Range.

Drainage Features. A narrow lake was dammed by moraines in the area between Sinyalak and Siruk Creeks (Pl. 3). The lake was at least 6 miles long, and its former extent coincides with a stretch of pronounced meanders along the Alatna River. Cutbanks expose about 20 feet of laminated silt and muck. Flood-plain deposits that stand about 35 feet above the present river surface near Siruk Creek may be relics of a higher river level graded to this lake.

Ground moraine just east of the mouth of Sinyalak Creek probably forced the lower part of the creek into a southwest-trending course, resulting in truncation of the eastern tip of the bedrock ridge between Sinyalak Creek and the Alatna River.

ITKILLIK GLACIATION

Drift in the Alatna Valley north of Siruk Creek is well-preserved relative to the more subdued glacial landforms farther south. Postglacial sediments above the drift are thin or absent, in contrast to the thick caps of peat and loess that typically cover till and outwash of Kobuk age. Ice limits in the Siruk Creek area appear to correlate with the lowest level of truncated bed-

TABLE 3. MEASURED SECTION OF TRENCH CUT INTO BLUFF ALONG SOUTHWESTERN SIDE OF ALATNA RIVER 7 MILES SOUTHEAST OF SINYALAK CREEK (PL. 3, LOCALITY 5)

Unit	Thickness (feet)	Lithology
A	5.5	Peat and silt; modified by root penetration and migration of iron oxide. The peat layers have been oxidized to reddish horizons that show open platy texture and contain spruce cones and needles.
B	1.0	Peat and clayey silt. Contains wood fragments that show pronounced flattening.
C	0.3	Intense iron oxide accumulation; decayed wood abundant. May represent former permafrost table.
D	5.0	Stratified peat and clayey silt. Contains wood fragments, usually as detritus in the silt. Wood shows no flattening and is very fresh in appearance. Three thin horizons of iron oxide concentration may represent rise of permafrost table as deposit accumulated.
E	2.5+	Laminated lacustrine sediment; silt and clay alternating with fine organic detritus containing twigs and other macroscopic plant remains. Laminae form apparent rhythmic couplets 5 to 10 mm thick. Wood near upper contact dates 8870 ± 460 years B.P. (GX0523). Base of unit extends below trench base.

rock spurs that flank the upper Alatna Valley (Pl. 2). Crests of these glacial facets attain a maximum altitude of about 4500 feet a few miles south of the present head of the Alatna drainage system, indicating a probable former ice divide there. The inferred glacier surface declines in altitude north and south of this area. It stood at about 2000 feet near the southern flank of the Brooks Range and farther down the Alatna approached moraines that extend into the valley near Siruk Creek.

This glaciation represents the last major episode of glacier erosion in the central Brooks Range and the final stage of ice advance into the lower Alatna Valley. It is assigned to the Itkillik glacial stage, named by Detterman (1953) and recognized throughout much of the north-central Brooks Range and adjacent foothills as the last major advance of late Pleistocene glaciers to positions at and beyond the range front (Detterman and others, 1958, 1963; Chapman and others, 1964; Porter, 1964, 1967). Porter has shown that the Itkillik Stage in the Anaktuvuk Valley is broadly comparable to the late Wisconsin and consisted of four substages.

Four distinct stillstands or readvances in the Alatna Valley during the Itkillik Glaciation are recognized. These appear broadly comparable to those of the Anaktuvuk area but will be described here using informal local names pending completion of current studies in the central Brooks Range. Most glacial deposits of Itkillik age in the Alatna Valley are concentrated in four arcuate belts that intersect the Alatna River near Siruk Creek, Chebanika Creek, the Helpmejack Lakes, and the southern range front (Pl. 5). Differences in postglacial modifications are relatively slight among these drift units and are far less than the contrast with features of the preceding Kobuk Glaciation.

Siruk Creek Drift

A belt of arcuate end moraines with superimposed knobs and ridges of ablation drift overlaps distinctly older glacial deposits north of Siruk Creek between the Alatna River and Norutak Lake (Pl. 5). The moraine belt trends southeast to near the mouth of Siruk Creek then curves upvalley to cross the Alatna River and form a drift embankment against the flank of a bedrock ridge farther to the northeast. These deposits define the maximum extent of ice during the Itkillik Glaciation.

Moraines and Drift. An end moraine up to 0.5 mile wide trends northeast from near the mouth of Siruk Creek to intersect the Alatna River (Pl. 5). Bedrock is exposed near the base of the moraine at its northeastern tip and probably has been a major factor in its preservation from postglacial stream erosion. Much of the moraine's surface is relatively smooth; other areas show undulating hummocks and swales or sharper knob-and-kettle topography. Most of the level till surfaces are covered by cottongrass tussocks and stunted spruce, indicators of poor subsurface drainage. *Sphagnum* forms thick insulating mats in many places, and permafrost commonly lies within 1 foot of the ground surface. Hummocky areas usually have 4 to 10 feet of relief, with broadly rounded and relatively well-drained hillocks underlain by sandy gravel and marshy swales underlain by till. A few larger kettle depressions and steep-sided kames are present across the moraine surface. Most kettles have been filled to form level marshes, but shallow water sometimes is present at their centers. The kames are underlain by poorly sorted gravel containing faceted and striated stones in a matrix of medium to coarse sand.

The northwestern flank of the moraine has been steepened by undercutting along the Alatna River. Till is exposed on the ridge flank, and lag concentrations of cobbles and boulders lie along the river edge. Slope movements have caused mobilization of the surface organic mat, which becomes segmented on upper parts of the slope and slides downward to accumulate as crumpled ridges. Till, exposed by displacement of the organic mat, is transported downslope by rain wash and mudflows to form lobes of detritus along the river. Finer components of the till have been washed away by the Alatna at the tips of these lobes, leaving lag deposits of subangular to rounded cobbles and boulders. Exposures along the slope exhibit bouldery till containing nonsorted stones in a clayey, sandy matrix. Few erratic boulders occur in the Alatna River downstream from the northeastern tip of the ridge, and lag deposits of cobbles and boulders at the ridge base terminate abruptly here.

Steep-sided kames, which rise a few tens of feet above bedrock or till surfaces, occur along the moraine near the Alatna River and extend to the north and southwest in a belt about 1 mile wide. Their hummocky, well-drained crests support open stands of birch, poplar, and white spruce above a ground cover dominated by the lichen *Cladonia*. Excavations show a few inches of eolian silt to fine sand above poorly sorted gravel, and slightly weathered granite boulders are scattered over most surfaces. The kames appear to coincide in position with the limit of Siruk Creek ice and with a minor recessional stand about 0.5 mile upvalley (Pl. 5).

Moraines farther west intersect Siruk Creek 3 to 5 miles southwest of its mouth. Boulders are scattered over the hummocky, well-drained surfaces of a series of short discontinuous ridges that stand about 75 feet above adjacent parts of Siruk Creek. Other moraines northeast of the Alatna River appear as a pair of ridges that connect groups of kames and extend farther north to form an embankment against the flank of bedrock uplands near the head of Sinyalak Creek.

Lateral Features. The moraines can be traced along both sides of the Alatna Valley into highlands near the Helpmejack Lakes. Further continuation of ice limits along the flanks of these highlands is indicated by distribution of erratic stones, ice-marginal drainage channels, overridden ridge flanks, ridge facets, and kame terraces.

An ice-scoured ridge on the flank of mountains west of the Helpmejack Lakes forms a pronounced flat-crested bench at an altitude of about 1700 feet. From a distance, the bench appears to have a smoothed overridden surface, with stoss-and-lee topography trending southward in the direction of ice flow. In detail, schist outcrops have a jagged appearance and frost shattering has eradicated most traces of their former abraded surfaces. Unaltered quartz pebbles and a few weathered granite cobbles are scattered across the bedrock. Possible kame terraces or lateral moraines cross valley mouths 2 to 4 miles southwest of the overridden bench and decrease in altitude from about 1600 feet near the Helpmejack Lakes to 1500 feet farther to the southwest.

Moraines east of the Alatna River extend into highlands northwest of Chebanika Creek. Marginal drainage channels, moraine or kame embankments, and ridge-facet crests increase in altitude northward and approach 1800 feet opposite the Helpmejack Lakes (Pl. 2).

Drainage Features. The Siruk Creek glacier extended into parts of the Siruk drainage, forcing the lower part of this stream into an ice-marginal course that intersects the Alatna River downvalley from its probable preglacial confluence. The present course of Siruk Creek lies partly beyond the outermost Siruk moraine and in part appears to trend between the outer moraine and its nearest recessional ridge.

About 3.5 miles southwest of its mouth, Siruk Creek begins to incise into graywacke bedrock. Immediately upstream from the incision, the creek flows at a gentle gradient between banks about 10 feet high that expose stratified silt probably of lacustrine origin. Flat marshy terrain covers the valley bottom for more than 9 miles southwest from this area and may indicate a former lake dammed by glacier ice (Pl. 3).

Chebanika Drift

A subsequent stillstand or minor readvance of the Alatna Valley glacier is marked by a prominent end moraine that intersects the Alatna River about 2 miles downvalley from Chebanika Creek. Outwash extends southeast from the moraine front as a nearly continuous terrace along the western side of the Alatna and can be traced downvalley for about 6 miles.

Moraines. A moraine nearly 1 mile wide trends south across Chebanika Creek and curves southwest into the center of the Alatna Valley (Pl. 5). The main body of the moraine stands about 60 feet high and has a poorly drained surface that contains a few kettle ponds. Three narrow ridges rise about 50 feet above the moraine surface and parallel its arcuate trend. The ridge crests have a hummocky appearance, and small kames stand in several places along their upvalley flanks. North of Chebanika Creek, the moraine extends toward the flank of the Alatna Valley and forms a drift embankment near the base of the valley wall. Near the valley center, the moraine has been eroded by the Alatna River, which presently flows through the Chebanika terminal zone with no change in gradient. The southwestern tip of the moraine segment east of the river has brushy unstable slopes, modified by slumps and mudflows that form lobes and fans of gray, stony clay along the river edge. A probable kame terrace composed of bedded sand and gravel stands about 35 feet high against the upvalley flank of the moraine there.

Discontinuous moraine segments west of the river curve toward the valley center, forming an arc that formerly extended to the moraine on the opposite side of the Alatna (Pl. 5). A moraine remnant near the river stands as a short ridge more than 50 feet high that bears a heavy brush cover indicating poor subsurface drainage. Sediments up to boulder size are exposed along the northern flank of the ridge. Subrounded gravels predominate on slopes where finer sediments have been removed by rain wash and clayey silt has been deposited along the base of the ridge flank. The Alatna flood plain here contains several well-drained areas that probably are underlain by sediment coarser than the usual flood-plain silt. These sites may contain lag deposits of cobbles and boulders formed during erosion of the moraine. Morainal ridges become more continuous away from the valley center and extend northwest into a probable drift embankment against bedrock.

Outwash. An outwash apron slopes downvalley from the Chebanika moraine into a terrace along the southwestern side of the Alatna River. Cutbanks immediately south of the moraine stand 50 feet high and expose a thin layer of peat over pitted outwash gravel. Downvalley, the outwash terrace commonly overlies graywacke at the river edge and wedges out laterally against bedrock ridge flanks to the southwest. The terrace surface is relatively well-drained, bearing open stands of birch and small spruce, but *Sphagnum* is dominant over the lichen *Cladonia* that covers younger outwash surfaces closer to the Brooks Range. The outwash terrace also is more dissected than younger outwash upvalley and is cut by broad gullies lined with dense willow and alder brush. River cuts exhibit well-sorted bouldery alluvium, and lag deposits along the southwestern bank of the Alatna contain cobbles and small boulders. Abandoned point bars of reworked outwash overlain by stratified peat and silt extend as much as 1 mile northeast from the terrace edge to the present course of the Alatna River.

The outwash terrace is continuous along the southwestern side of the Alatna flood plain to within 0.5 mile of the mouth of Siruk Creek (Pl. 5), maintaining through this distance a height of 30 to 40 feet and a well-drained surface marginal to bedrock ridges farther southwest. Near the mouth of Siruk Creek, the terrace appears to merge with a broad, nearly level surface that stands about 30 feet above the Alatna River. Texture and sorting of the outwash show no pronounced change upvalley from Siruk Creek, indicating that it probably was deposited while ice retreated from the Siruk Creek position to the Chebanika area.

Drainage Changes. Possible ponding of Alatna drainage behind the Chebanika moraine is indicated by abundant organic silt in the river banks and flood plain upstream from the moraine and by an extensive, level, poorly drained surface west of the river (Pl. 5). This surface contains numerous thaw lakes, irregular in shape and distribution, that contrast with oxbow patterns marking the recent Alatna flood plain.

Drainage along Siruk Creek probably was graded to the 30-foot outwash surface near its mouth. Bouldery gravel 2 to 3 miles up Siruk Creek rests on bedrock banks as much as 44 feet high. The gravel contains faceted and striated stones that include several exotic rock types. Probably they were derived from the ice front during recession from the Siruk Creek stand and were transported along channels through the inner Siruk Creek moraine.

Helpmejack Drift

Massive ice-contact stratified drift surrounds the Helpmejack Lakes and merges southward with moraines and outwash deposited during a major still-stand or readvance of the Alatna Valley glacier during the Itkillik Glaciation.

Extent and complexity of the drift suggest that glacier ice occupied a terminal position in the Helpmejack Lakes area during a relatively long span of time.

Terminal Moraines and Related Drift. The southernmost end moraines of the Helpmejack area are best exposed west of the Alatna River 7 miles downvalley from the mouth of Helpmejack Creek. Two parallel ridges stand 100 feet above the Alatna flood plain and trend northwest away from the valley center (Fig. 4). The flanks and hummocky crests of the moraines exhibit sand and gravel in all natural exposures; a trench dug into the crest of the outermost moraine exposed 5 feet of ice-contact sediments above unweathered, gravelly till. The moraines curve eastward close to the valley center and are aligned with a series of drift ridges and embankments against highlands on the opposite side of the Alatna River. The Alatna flows past the moraine belt with little change in gradient, but immediately downstream it deepens and passes through a narrow canyon flanked by bedrock ridges.

Stratified drift (Fig. 4, sd_1) lies between the moraine front and the bedrock ridge that forms the western flank of the canyon. Depositional features include flat-crested ridges and mounds and oblong bodies that resemble moulin and crevasse fillings. Most rise only a few tens of feet above the general surface of the deposit, which forms an embankment against the northern tip of the bedrock ridge. Fluvial sand and gravel are exposed in most cutbanks, and well-sorted pond deposits appear in several places. Most of the sediment probably was deposited in contact with stagnant glacier ice, but some may have been associated with ice blocks in melt-water channels. The deposits may represent either an early event of Helpmejack time or a late stagnation phase of the Chebanika episode.

An outwash terrace (Fig. 4, ow_2) extends south into the canyon from the moraine ridges west of the Alatna River. The terrace forms a relatively level, well-drained surface west of the Alatna and is bordered farther west by the ice-contact deposits and the bedrock-ridge flank. The terrace stands 80 feet above the river near the moraine front but slopes southward and stands only 40 feet above the Alatna 2 miles downvalley. Sand and gravel were found in all excavations. Bedrock exposures along the Alatna reach a maximum height of 50 feet near the southern end of the canyon but average 30 feet where exposed beneath outwash closer to the Helpmejack terminal moraine. The outwash, therefore, forms a wedge that decreases in thickness from about 50 feet near the moraines to 10 feet about 2 miles downvalley. It appears distinctly older than outwash near the Helpmejack Lakes because of its well-developed soil. Excavations exposed podzolic soils with eluviated horizons averaging 1.5 inches thick and with clay accumulation in the B horizon. Postglacial stream channels incised into the outwash terrace bear dense willow and alder cover above organic silt. Most are graded to the bedrock surface, which forms a bench along the river edge.

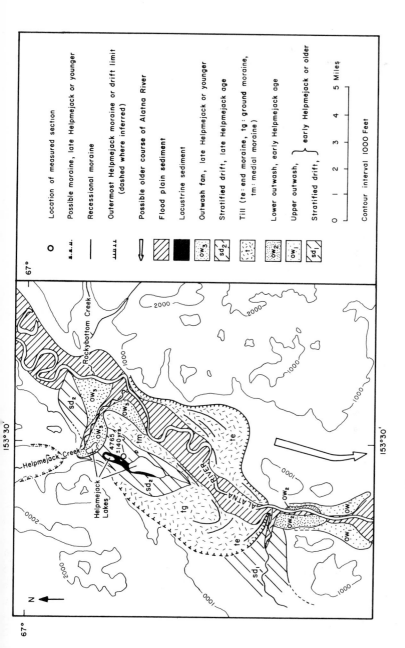

Figure 4. Generalized surficial geology, Helpmejack Lakes area.

Legend

○   Location of measured section

Possible moraine, late Helpmejack or younger

Recessional moraine

Outermost Helpmejack moraine or drift limit (dashed where inferred)

Possible older course of Alatna River

Flood plain sediment

Lacustrine sediment

ow₃   Outwash fan, late Helpmejack age

sd₂   Stratified drift, late Helpmejack age

Till (te: end moraine, tg: ground moraine, tm: medial moraine)

ow₂   Lower outwash, early Helpmejack age

ow₁   Upper outwash, } early Helpmejack or older

sd₁   Stratified drift,

0 1 2 3 4 5 Miles

Contour interval 1000 Feet

Remnants of two probable outwash terraces extend along the eastern side of the river through the canyon area. The lower lies at about the same level as the outwash terrace west of the Alatna, but it has been more extensively eroded and is preserved only as a discontinuous line of well-drained benches and elongate knobs along the inner flank of the bedrock ridge east of the canyon. The higher line of terrace remnants is identical in appearance and is matched west of the river by a set of similar elongate deposits at about the same altitude. The higher terrace remnants both east and west of the river extend about 1 mile downvalley beyond the lower outwash terrace, then spread laterally into a fan-shaped deposit south of the canyon (Fig. 4, ow_1). Lateral erosion by the Alatna River beyond the confines of the bedrock canyon has removed much of the fan.

Recessional Drift. Moraines between the Helpmejack Lakes and the terminal zone to the south represent two phases of ice recession. An initial stage of fluctuating glacier retreat is indicated by narrow moraine ridges east of the Alatna River that trend toward the valley center from a drift embankment at the base of the highlands opposite the Helpmejack Lakes. These ridges appear to be related to similar moraine segments west of the Alatna near the terminal zone. A later stage of ice stagnation followed division of the Helpmejack glacier into two lobes separated by bedrock and medial moraine between the Helpmejack Lakes and the Alatna River.

Deposits of ice-contact stratified drift surround the southernmost Helpmejack Lakes (Pl. 6, fig. 1; Fig. 4, sd_2). The drift forms terraces against the inner flanks of moraines and bedrock ridges to the east and west and extends across the intervening area as a series of irregular to arcuate knobs and ridges. Most of the features appear to be kames and kame terraces, deposited in contact with the margins of large stagnant-ice blocks detached from the active portion of the thinning Helpmejack glacier. The sediments are very well drained and are marked by steep but stable flanks and by relatively sparse plant assemblages dominated by *Cladonia*, aspen, and white birch.

The higher ridges that border the lakes to the east and west are underlain by a veneer of impermeable clayey till above conglomerate bedrock and bear a relatively dense cover of spruce and *Sphagnum*. The ridge between the lakes and the Alatna River was overridden by glacier ice during early Helpmejack time, and its southern tip probably was extended southward by the deposition of medial moraine (Fig. 4, tm). The ridge subsequently emerged above the surface of the thinning glacier, and kame terraces were deposited against its western flank when ice became stagnant in the Helpmejack Lakes area. Low bedrock ridges west of the lakes were overridden to more than 1000 feet altitude during early phases of the Helpmejack Stade, and granite erratics are scattered along their crests. Younger moraines and deposits of stratified drift stand at about 800 feet altitude along the eastern flanks of the ridges. These

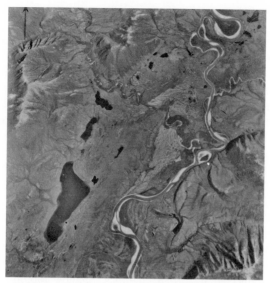

Figure 1. Drift around Helpmejack Lakes. Scale decreases northward (direction of arrow) from lower margin of photograph.

Figure 2. End moraines south of Iniakuk Lake, September 1963. Contrasts between light-toned aspen-*Cladonia* and dark-toned spruce-*Sphagnum* associations are particularly striking during autumn color change, when aspen leaves turn brilliant yellow. Mapping of surficial geology by air reconnaissance and photography is most effective during this interval.

ITKILLIK DRIFT, LOWER ALATNA VALLEY

deposits are relatively well drained and include both isolated kames and broader kame terraces.

Stratified drift (Fig. 4, sd$_2$), probably of combined outwash and ice-contact origin, forms a broad surface south of Helpmejack Creek 2 miles above its mouth. The deposit stands about 115 feet above the creek and extends south and west to the drift ridges around the lower Helpmejack lakes. Much of the sediment may have been deposited by outwash streams that flowed from the Alatna Valley or Helpmejack Creek into the area occupied by large stagnant ice blocks. The deposit has been dissected along its northern end by Helpmejack Creek, forming a scalloped bank 60 feet high above a younger outwash terrace.

A lower surface of probable ice-contact origin, also shown on Figure 4 as sd$_2$, extends north from Helpmejack Creek along the base of highlands west of the Alatna River. It appears relatively level in places but elsewhere is interrupted by straight or irregular ridges up to 30 feet high and by lines of kettle ponds connected by broad channels. Shallow pits and trenches dug along ridge crests exposed well-sorted sand above fluvial gravel. One group of straight-sided ridges and chains of kettles, which trends southeast toward the Alatna River, may represent a set of crevasses that initially were filled with sediment and later became the locus of melt-water channels across the area. Level sites near the northeastern margin of the area are poorly drained and generally bear a cover of stunted spruce and deep *Sphagnum*. River banks expose lacustrine sand above fluvial gravel; the sand contains horizontal beds of clay and silt and possibly was deposited in a lake formed behind the main drift body.

Drainage Features. Two major drainage diversions occurred during Helpmejack time. Helpmejack Creek, which probably once flowed south to join the Alatna River below the lakes, was diverted eastward by massive glacial deposits in the present lake area. Farther south, the Alatna River probably was diverted into its modern course through the bedrock canyon below the Helpmejack terminus by a drift embankment east of the present valley center. This deposit, possibly formed by late Chebanika as well as early Helpmejack glaciation, blocked a broad valley that extends south nearly to Chebanika Creek (Fig. 4). Postglacial dissection of bedrock along these new drainage courses has been on the order of 50 feet in the canyon area and 60 feet where Helpmejack Creek cut through the ridge between the lakes and the Alatna River.

Several of the Helpmejack lakes show evidence of formerly higher levels. A conspicuous shoreline around the southernmost lake stands about 30 feet above the present lake surface. Steep kame flanks above this height contrast with gentle slopes below that are underlain by lacustrine sand. A dissected surface 20 to 30 feet high between the two southernmost lakes is underlain by poorly drained bog and lacustrine sediments that bear a cover of spruce, dwarf birch, sedges, and *Sphagnum*.

Helpmejack Outwash Fan. Outwash along the lower course of Helpmejack Creek extends east past the lake area and spreads into a fan-shaped body that intersects the Alatna River (Fig. 4, ow$_3$). The fan transects ice-contact stratified drift and, around the lakes, is clearly younger than these deposits. Its relation to subsequent glacial events is uncertain, however.

North of the lakes, the outwash forms a broad terrace 45 feet high that lies between Helpmejack Creek and the scalloped flank of older outwash deposited during ice stagnation in the lake area. The terrace extends northeast up Helpmejack Creek to the mouth of its bedrock valley. Other deposits about 2 miles up this valley appear continuous with outwash related to range-front moraines farther north. Outwash in the lakes area and near the southern margin of the Brooks Range may form a continuous unit or may be separated by an intervening, poorly defined moraine (Fig. 4).

North of Helpmejack Creek, the outwash slopes east and northeast toward the Alatna River. Shallow braided channels that parallel this slope form a distinctive pattern across the surface of the fan. (Pl. 6, fig. 1). The outwash is separated from older ice-contact deposits by a low bank that appears to have been cut by stream erosion. East of this bank and close to the Alatna River, the outwash surface is interrupted by a deep broad channel, pitted with kettle ponds, that extends southeast from the older drift. Braided patterns on the outwash appear to cross this feature, as individual shallow drainage lines that terminate at the western bank of the channel can be extended visually into the outwash beyond its eastern bank. The kettle-filled channel must have still contained ice blocks at the time the outwash fan was formed. Outwash probably was deposited across these blocks, but the continuity of the fan surface later was broken by melting of the buried ice.

The outwash forms a bank 55 feet high along the western side of the Alatna just above Helpmejack Creek, and outwash exposures extend 2 miles farther north along the river. The bank decreases in height northward and averages 35 to 40 feet near its upvalley limit. Alluvium becomes less coarse to the north: small boulders become rare, cobbles decrease in abundance relative to pebbles, and sand becomes dominant. Small to medium boulders are common in lag deposits along the river edge opposite Rockybottom Creek, however, indicating that till may underlie outwash here.

South of Helpmejack Creek and close to its mouth, the braided channels on the outwash surface trend downvalley parallel to the present river course. An outwash bank west of the Alatna below the Helpmejack confluence averages 40 feet high; it exposes rounded to subrounded, fairly well sorted alluvium of sand to cobble size. Lag deposits along the river bank consist mainly of pebbles and cobbles with a few small boulders.

The outwash fan may be related to either a very late phase of Helpmejack time or to a separate stade intermediate between those represented by the Helpmejack and range-front drifts. Fan deposition may be related to a former glacier terminus in Helpmejack Valley 6 miles south of the outermost range-front moraines, ice blocks persisting in the Helpmejack drift appear to have been overridden by the outwash, and the northeastern margin of the fan may have been deposited against an ice front in the Rockybottom Creek area.

Range-Front Drift

Terminal moraines and outwash are located at the southern flank of the Brooks Range along the Alatna Valley and two glacial distributaries (Fig. 5). The distributary valleys, now occupied by Tobuk Creek and Iniakuk River, received overflow ice that crossed low divides east of the main Alatna Valley glacier. Drift along the Alatna River extends from the range front south to Malemute Fork and upper parts of Helpmejack Creek. Moraines east of the Alatna enclose Iniakuk Lake and cross the valley now occupied by Iniakuk River. These deposits represent the latest stade of the Itkillik Glaciation in the lower Alatna Valley.[1]

Drift near Iniakuk Lake. An arcuate belt of terminal moraines and ice-contact stratified drift extends about 4 miles between bedrock ridges that border the valley of Tobuk Creek (Fig. 5). The moraines enclose Iniakuk Lake and, farther south, extend into the valley of Malemute Fork. The moraine area is 0.5 mile wide southwest of Iniakuk Lake, but its width increases to more than 1 mile near its eastern end.

Three principal moraine ridges cross the valley, and at least two discontinuous moraines are present farther north near the bases of the bedrock-ridge flanks. Several melt-water channels have breached the moraines (Fig. 5), but most of the outwash transported south through the drift belt has been reworked subsequently by Malemute Fork. The moraines form steep-sided prominent ridges that stand as much as 75 feet high where cut by the Iniakuk Lake outlet stream. Their crests usually are irregular, bearing well-preserved ridges and hummocks of ablation drift. Aspen generally is the only tree species along the ridges, and nearly continuous mats of *Cladonia* dominate the underlying ground cover. Depressions between the moraines are occupied by stands of black spruce and by *Sphagnum* bogs or marshy areas of shallow open water (Pl. 6, fig. 2).

A section through the outer moraine is exposed where this ridge is breached by the outlet stream from Iniakuk Lake. Six inches of well-sorted eolian sand overlie 55 feet of gray till that is highly compact and stands as a

---

[1] Their limits approximately coincide with a previously inferred boundary of "glacial advances of late Pleistocene age" (Coulter and others, 1965).

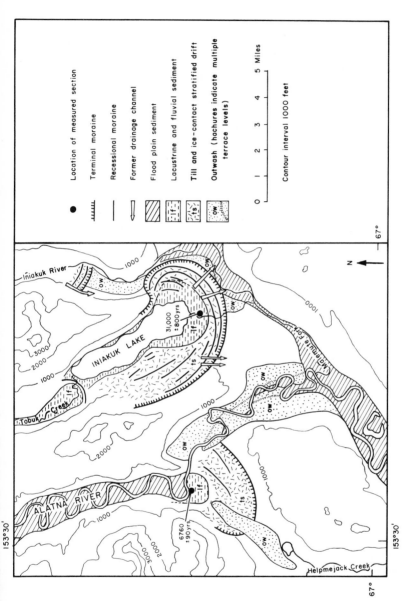

Figure 5. Generalized surficial geology, Alatna Valley at range front.

steep bank. The till is composed of nonsorted subangular to rounded sediment of clay to boulder size. Granite and quartzite stones in the section are essentially unaltered, but limestone, schist, and mafic igneous rocks are conspicuously weathered. A similar exposure lies 1 mile to the northeast, where the downvalley flank of the outer moraine has been cut by a meander loop of Malemute Fork. A nearly vertical bank 60 feet high exposes bouldery till with local sorting. Pits excavated along moraine crests showed similar tills, modified in some areas by removal of the finer sized components by melt water. Near the western end of the moraine belt, large subangular fragments of schist and quartz become abundant in the till. These probably were derived from the adjacent bedrock ridge that lies between Iniakuk Lake and the Alatna River.

An outwash apron extends south from the moraine belt near the valley center. Outwash is most conspicuous immediately east of the outlet stream, where it forms a well-drained terrace that slopes southward to the flood plain of Malemute Fork. To the north, the outwash terminates against the downvalley flank of the outermost moraine. The terrace stands 36 feet high near the moraine and thins southward to form a tapering wedge above the northern flank of schist bedrock that intersects the outlet stream near its mouth. The outwash surface is level and well-drained; it bears a cover of *Cladonia* beneath open stands of birch, spruce, and aspen. Soil development of the outwash is similar to that on the moraines, but the parent material is better rounded and lacks sediment finer than medium sand. Many cobbles and small boulders are present in outwash near the moraines, but farther south small cobbles generally constitute the largest stones. Sections near Malemute Fork typically expose 8 inches of fine, silty sand above 1 to 3 feet of fluvial sand and gravel. The bedrock beneath the outwash is highly weathered at the contact, and frost-shattered schist fragments are mixed with the lower few inches of alluvium.

Lateral moraines, ice-marginal drainage channels, and concentrations of erratic boulders extend north from the moraine belt along the inside flanks of the schist ridges that border Iniakuk Lake. The upper limits of these marginal features rise from 1100 feet altitude near the southern end of the lake to at least 1500 feet 5 miles to the north.

Drift along Iniakuk River. Two well-defined end moraines cross the valley of Iniakuk River at the northern limit of a pitted outwash train (Fig. 5). Two sharp-crested ridges of well-drained ablation drift are superimposed on the surface of the southern moraine and stand close together at heights of 30 to 40 feet above the outwash. The northern moraine has a single crest and trends parallel to the southern drift ridges. The moraines are separated by a depression 0.5 mile wide that contains low ridges and linear bogs parallel in trend to the adjacent moraine crests. Melt-water channels east and west of the moraines extend along the bases of the valley walls. The eastern channel subsequently has been incised about 20 feet into schist near the upvalley moraine and contains the modern channel of Iniakuk River. The western channel, which contains a series of bogs and shallow ponds, rises toward the north and probably is related to ice-marginal drainage features that trend southeast from upper Tobuk Creek into the valley of Iniakuk River.

Close to the moraines, the outwash stands as a terrace 30 feet above Iniakuk River. Cobbles and small boulders dominate the lower part of bank exposures; sand and pebbles become more abundant closer to the terrace surface. Stones in the outwash are rounded to subrounded, stratified, and cemented by calcium carbonate in places. The upper 10 to 14 feet have been oxidized. Farther downvalley, the outwash terrace is lower and boulders are less abundant.

Drift along the Alatna River. Arcuate morainal ridges extend southwest from the Alatna River at the range front, reach their southernmost position near upper Helpmejack Creek, then trend north to the western flank of the Alatna Valley (Fig. 5). To the east, the moraines have been breached by an extensive outwash train and by later erosion of the Alatna River. To the west, they have been dissected by melt-water streams whose deposits extend southwest to Helpmejack Creek. The moraines are far less conspicuous than the many kames and irregular ridges that mantle their crests and flanks. These superimposed bodies usually are steep-sided and well-drained and are accentuated by the distinctive light tone of aspen-*Cladonia* assemblages. West of the Helpmejack outwash train, the position of the glacier terminus is represented by kame terraces against the base of the mountain front. These stand at an altitude of about 900 feet in the terminal zone and rise northward into the Brooks Range to join ice-marginal drainage channels along the flank of the Alatna Valley. To the east, the moraines are represented only by boulder concentrations along the present course of the Alatna River. Boulders first appear where the innermost moraine intersects the river and boulders and cobbles are common in lag concentrations along the river banks for several miles downstream.

An outwash train east of the moraine ridges crosses the terminal belt and extends about 6 miles south along the present course of the Alatna River (Fig. 5). The outwash has been extensively reworked by the Alatna and its tributaries, forming several sets of terraces near the terminal zone. The upper terrace stands about 75 feet above the present river level and is slightly lower than the moraines that extend to the southwest. The northern segment of the terrace is about 1 mile wide and extends between the Alatna River to the west and ice-marginal drainage channels along the valley wall to the east. Braided channels on the outwash surface suggest flow directions toward the southwest in some places, indicating that part of the deposit was formed by drainage from the ice-marginal channels. Farther south, this outwash surface extends downvalley past the mouth of Malemute Fork and covers extensive areas between the valley walls and the margins of the Alatna meander belt. Outwash banks along the Alatna River are lower downvalley and stand only 45 feet high 3 miles south of the terminal zone. Cutbanks along the Alatna close to the moraines expose stratified outwash, ranging in size from sand to small boulders, which has been cemented by calcium carbonate in places. Boulders become rare in outwash farther south, and exposures seldom contain stones larger than coarse cobbles.

A discontinuous terrace of reworked outwash extends south along the Alatna River to a position near the mouth of Malemute Fork. This terrace usually is best developed near the edge of the meander belt and appears related to early stages in the extension of present meander loops across the Alatna flood plain. The terrace stands about 50 feet high close to the moraine belt and decreases in height downvalley to about 30 feet close to Malemute Fork. A cutbank along the Alatna 1.5 miles north of Malemute Fork exposes the contact between the older and the reworked outwash. The upper terrace stands 45 feet high, is composed of unweathered sediment, and is oxidized only at the surface. The reworked outwash stands 15 feet lower, and the slope between the two surfaces has a concentration of cobbles and small boulders similar to the lag deposits along the present river edge. The reworked outwash has the same general composition as the parent body, but it has a greater percentage of sand and is more discolored by oxides and organic material.

Outwash near the western end of the moraine belt extends 4 miles from the Alatna River to Helpmejack Creek, and discontinuous remnants are preserved along Helpmejack Creek for 3 miles farther south. The outwash train has breached the end-moraine belt, and morainal ridges terminate abruptly at the eastern margin of its channeled surface. Farther south, the level outwash surface is channeled in a shallow braided network that indicates flow directions toward the southwest. North of the moraine belt, the outwash train has a more irregular surface, with deeper channels between well-drained kames and short ridges of ablation drift. A concentration of kettle ponds at the boundary between the two contrasting segments of the outwash trains may mark the position of the glacier front during part of the outwash deposition.

Drainage Features near Iniakuk Lake. Iniakuk Lake occupies an elongate basin that probably was preserved from late-glacial filling by the presence of stagnant ice blocks. Soundings show that the lake is more than 80 feet deep. The basin therefore extends well below the level of the schist barrier crossed by the outlet stream and probably resulted from glacial overdeepening near the valley mouth.

The southern shore of the lake rises to an undulating surface, 15 feet above water level, which extends inland about 0.25 mile as a series of ice-shoved ramparts (Peterson, 1965) parallel to the present shoreline. Older lacustrine deposits, at heights up to 30 feet, form a series of ice-shoved ramparts along the eastern and western shores. South of the lake, a nearly level surface 30 feet high extends to the inner flank of the moraine belt. Banks along the outlet stream expose fluvial sand and gravel above probable lacustrine sand. Bedding in the fluvial horizons usually dips east, indicating that much of the sediment may have been transported along the base of the ridge flank west of the present lake margin. The 30-foot lake stand may have been controlled by the schist ridge across which the outlet stream flows. This ridge has been breached by a narrow gorge as much as 33 feet deep.

Traces of a shoreline, marked by beach deposits of well-rounded pebbles and coarse sand, extend along the eastern and western sides of the lake at a height of about 50 feet. The moraine belt south of the lake, where intersected by Malemute Fork and the outlet stream, has been dissected to form nearly vertical banks of till 55 to 60 feet high. More gentle slopes rise above these banks and extend back to the moraine crests. The change in slope at the tops of the banks may be related to a higher drainage level, graded perhaps to the highest outwash terrace along the Alatna River, that controlled the height of Iniakuk Lake during early stages of deglaciation.

Drainage Features along the Alatna River. Drift aggraded the area southwest of the present Alatna River near the mountain front and caused eastward displacement of the river course. The Alatna was forced into a new position across a bedrock ridge into which it has entrenched a narrow gorge about 50 feet deep. Prior to drainage displacement, the Alatna River may have flowed southeast through the moraine area 1 mile west of its present course or possibly flowed south through the valley of Helpmejack Creek. The massive outwash fan at the mouth of Helpmejack Creek indicates considerable meltwater discharge in this area during or following late Helpmejack time, but subsequent dissection has been concentrated mainly along the present course of the Alatna River.

Deglaciation was accompanied by outwash deposition along both the Helpmejack drainage and the Alatna River. The outwash trains stand at nearly comparable altitudes, but greater postglacial stream dissection of the eastern body resulted from the additional discharge contributed by Malemute Fork. Most of this outwash was eroded and redeposited at a lower level controlled in part by the schist ridge across the eastern end of the Alatna moraine belt. A later halt in downcutting through the ridge occurred at a level about 25 feet above the present river. Flood-plain and lacustrine sediments rise to this height north of the ridge, and bedrock remnants stand about 25 feet high downvalley from it.

CORRELATION AND DATING OF THE GLACIAL SEQUENCE

Lower Alatna Valley

Subdivisions and suggested correlations of the glacial record are summarized in Table 4. The positions of moraines and associated drift sheets are the primary basis for differentiating seven glacial events in the lower Alatna Valley and assigning relative ages to them. Contrasting degrees of postglacial wastage and modification confirm the relative ages indicated by moraine positions and suggest three principal age groups for the glacial deposits. The oldest glaciation and the subsequent Kobuk and Itkillik Glaciations are represented also by distinct levels of faceted bedrock spurs within the upper Alatna Valley. Correlations with the standard North American glacial sequence are based on radiocarbon dates from the Alatna Valley and comparison with dated events elsewhere in northern Alaska.

Drift of the oldest glaciation lies well beyond the mouth of the Alatna Valley and is not exposed in any of the bluffs along adjacent parts of the Koyukuk River. Its study therefore was limited to aerial reconnaissance and interpretation of aerial photographs during my investigations of the Alatna Valley region. Although little is known of the physical characteristics of the drift, its distribution can be inferred from outer limits of hummocky terrain and deranged drainage and from upper limits of glacier and melt-water activity. These limits outline the position of a piedmont ice lobe that occupied an area of at least 2000 square miles within the Koyukuk Lowlands. The great extent of this glacier through the lowlands south of the Brooks Range exceeds that of any known event of Wisconsin age in northern Alaska. A piedmont lobe of comparable size, fed by glaciers in the Kobuk and Noatak Valleys, has been assigned an Illinoian age on the basis of stratigraphy and radiometric dates in the Kotzebue Sound area (McCulloch and others, 1965).

Drift assigned to the subsequent Kobuk Glaciation is marked by broad subdued morainal ridges, more hummocky terrain, and preservation of ice-marginal stream courses. Two stades are defined on the basis of separate moraine belts south of the Koyukuk and 25 miles farther north within the Alatna Valley. Drift of the recessional stade clearly is younger than that of the maximum advance because of its position closer to ice sources within the Brooks Range. Eolian and bog sediments typically are less than 6 feet deep above drift of the recessional stade, in contrast to thicknesses up to about 50 feet over till and outwash of the maximum advance. Both stades are considered subdivisions of the same glaciation because of similarities in morphology, stratigraphy, and postglacial modifications; drift near the western margin of the lower Alatna cannot be separated into distinct stadial components. Drift of Kobuk age is distinguished from the older deposits in being far more modified in appearance, and glacial facets of Kobuk and pre-Kobuk age occur at

TABLE 4. ALATNA VALLEY GLACIAL SEQUENCE AND CORRELATION WITH GLACIAL EVENTS IN ADJACENT REGIONS

Standard North American sequence	Alatna Valley		Northern Brooks Range (Detterman and others, 1958)	Anaktuvuk Valley (Porter, 1964, 1967)		Kobuk Valley (Fernald, 1964)
Neoglacia-tion	Valley-head advances		Fan Mountain	Fan Mountain II Fan Mountain I		Modern advances
			Alapah Mountain	Alapah Mountain		Ulaneak Creek
Late Wisconsin	Itkillik	Range front	Echooka		Anivik Lake	Walker Lake
		Helpmejack	Itkillik	Itkillik	Antler Valley	Ambler
		Chebanika			Anayaknaurak	
		Siruk Creek			Banded Mountain	
Early Wisconsin	Kobuk	Late stade	Sagavanirktok?			Kobuk
		Early stade				
Illinoian	Oldest glaciation		Anaktuvuk?			Earliest glaciation

separate levels in the upper Alatna Valley. Separation of the Kobuk Glaciation from the later Itkillik Glaciation is based on (1) differences in modification of drift, (2) distinct drift borders near the western side of the lower Alatna Valley, (3) altitudes of facet crests within the Brooks Range, and (4) differences in late-glacial and postglacial eolian deposition, organic accumulation, and drainage development. Till of the maximum advance and basal organic units above it are older than late Wisconsin on the basis of infinite radiocarbon dates obtained from organic detritus exposed in bluffs along the Koyukuk River. The Kobuk Glaciation may correlate with early Wisconsin advances elsewhere in Alaska (for example, Péwé and others, 1965; Sainsbury, 1967). An alternate correlation with glacial events of late Illinoian age is considered less likely because no evidence is present in the bluff exposures for significant weathering, soil formation, or permafrost recession associated with a major interglacial stage.

Drift of Itkillik age is less modified, and well-drained deposits such as kames and ablation moraines stand as sharp-crested, steep-sided knobs and ridges. Late-glacial and postglacial deposits usually consist of less than 1 foot of eolian silt or fine sand, in marked contrast to the thicker deposits above older drift. The extensive *Siruk Creek* moraine complex and related glacial facets farther north suggest a relatively long initial episode of glacier advance and active erosion. Water-washed sediments that include outwash, kame deposits, and stratified drift of ablation moraines are well-preserved, but bodies of nonsorted drift are more modified. The *Chebanika* moraine stands in relatively sharp relief and shows better preservation of minor irregularities on its surface. Chebanika deposits are less extensive and complex than those around Siruk Creek and may represent a relatively brief episode that followed the much longer Siruk Creek event. The *Helpmejack* drift also represents a long period of ice activity during the Itkillik Glaciation. It is comparable to the Siruk Creek drift in extent and complexity of glacial deposits and, like the older event, is associated with modifications of valley walls in the upper Alatna drainage. Helpmejack drift includes extensive stratified deposits associated with stagnant glacier ice; the former ice surface in the upper Alatna Valley is marked by altitudinal limits of nivation and melt-water erosion rather than those of glacial abrasion (Pl. 2). The *range-front* moraines form relatively narrow belts that lack the extent and complexity of the Helpmejack and Siruk Creek drift sheets. Moraines as well as kames and outwash deposits have been little modified by postglacial weathering and mass movements and have been extensively eroded only where large volumes of melt water flowed during glacial recession.

Eight radiocarbon dates were obtained from organic samples within the lower Alatna Valley and adjacent parts of the Koyukuk (Table 5). *Samples GXO524* and *UW-85*, from the Koyukuk bluff exposures, yielded infinite dates on sediments associated with the maximum advance of the Kobuk Glaciation. Sample GXO524, obtained near the base of the till at locality 2, was reported to have no radiocarbon activity above that of the normal background count (Geochron Laboratories, written commun.). Sample UW-85, from organic deposits immediately above the till at locality 1, exhibited sufficient activity to indicate that its age may lie within the range datable by radiocarbon-enrichment techniques (A. W. Fairhall, oral commun.).

Sample GXO523, from lacustrine sediments at locality 5 near Sinyalak Creek, was obtained from a trench that did not penetrate far into the lacustrine section. All but the upper few feet of the sediments here are masked by refrozen colluvium along the face of the river bluff, and undisturbed samples from basal parts of the unit can be recovered only by deep excavation into permafrost. The sample may be from a local thaw lake, and its age of 8870

TABLE 5. RADIOCARBON AGE DETERMINATIONS, LOWER ALATNA VALLEY AND
KOYUKUK BLUFFS

Sample no. and laboratory	Radiocarbon years B.P.	Location, material dated, and stratigraphic position
UW-85 Univ. Wash.	>38,000	Koyukuk bluffs, locality 1 (Pl. 3 and Table 1.) Peat at base of eolian and organic deposits above till.
GX0524 Geochron	>34,300	Koyukuk bluffs, locality 2 (Pl. 3 and Table 1). Small wood fragments within till near base of exposure.
GX0523 Geochron	8,870 ± 460	Locality 5 (Pl. 3 and Table 3). Wood from lacustrine sediments above drift; recessional stade of Kobuk Glaciation.
UW-83 Univ. Wash.	8,990 ± 80	Itkillik drift sheet; outwash downvalley from Chebanika moraine (Pl. 5). Base of thin silt unit above outwash.
GX0521 Geochron	8,205 ± 195	Itkillik drift sheet; kettle fill in outwash downvalley from Chebanika moraine (Pl. 5). Wood from 9 feet deep just above upper surface of outwash.
GX0522 Geochron	4,765 ± 140	Itkillik drift sheet. Wood from lacustrine deposits in area dammed by Helpmejack drift (Fig. 4 and Table 6).
UW-84 Univ. Wash.	6,760 ± 90	Itkillik drift sheet. Wood from lacustrine sediments in area dammed by end moraine at range front (Fig. 5 and Table 6).
W-1427 U.S. Geol. Survey	31,000 ± 800	Itkillik drift sheet. Small wood fragments from lacustrine sediments in area dammed by end moraine at range front (Fig. 5 and Table 6).

± 460 years B.P. probably does not provide a closely limiting date for the recessional stade of the Kobuk Glaciation.

Samples UW-83 and GXO521, dated 8990 ± 80 and 8205 ± 195 years B.P., were obtained from organic deposits above outwash related to the Chebanika moraine (Pl. 5). The outwash may be much older than the overlying organic sediments, which were obtained from limited exposures along cutbanks of the Alatna River and Siruk Creek. The outwash surface also could have remained an environment unfavorable for plant recolonization for some time after glacier recession from the Chebanika moraine. Establishment of extensive plant cover on much of the drift of Itkillik age might possibly be contemporaneous with a pronounced warming episode about 10,000 to 8300 years ago reported from northwestern Alaska (McCulloch and Hopkins, 1966), but direct evidence for this event has not yet been found in the Alatna Valley.

Samples GXO522 and *UW-84*, from lacustrine sediments confined behind drift barriers at Helpmejack Lakes and at the southern margin of the Brooks Range, are related to postglacial events in the region's drainage history. The date of 4765 ± 140 years B.P. was obtained from wood near the base of a peat horizon in lacustrine sediments at the Helpmejack area (Fig. 4; Table 6, section 1). The peat may have formed during an interval of lower water table associated with permafrost recession during the hypsithermal interval. The 6760-year-old sample from the Alatna Valley at the southern margin of the range dates an episode of filling by alluvium and lacustrine deposits within a basin dammed by end moraines and related drift (Fig. 5, unit lf; Table 6, section 2). Deposition postdated glacier recession from the moraine belt and predated final downcutting of the Alatna River to its modern level through the gorge at the southern margin of the range. About half of the 50 feet of bedrock incision apparently had been completed at the time the sediments were deposited; an unknown amount of this may have occurred subglacially before ice receded from the terminal zone.

Sample W-1427, from sediments deposited during deglaciation of the Iniakuk Lake area, yielded a seemingly anomalous age of 31,000 ± 800 years B.P. (Fig. 5; Table 6, section 3). If correct, this date must indicate that interglacial organic debris was buried by the initial glacier advance to Siruk Creek, remained in a protected site throughout the Itkillik Glaciation, and was redeposited after glacier recession from the Iniakuk Lake moraines.

Although radiocarbon dates provide little evidence for precise ages of the drift units, they do confirm some of the inferred stratigraphic subdivisions and age correlations of glacial deposits in the lower Alatna Valley. A late Wisconsin age for the Itkillik Glaciation is supported by limiting dates of about 6700 to 9000 years on three postglacial samples (GXO521, UW-83, and UW-84); the 31,000-year-old wood fragments from Iniakuk Lake (W-1427) may provide a valid limiting date for influx of Itkillik ice into the lower Alatna Valley. Drift exposed in bluffs along the Koyukuk River clearly is older than late Wisconsin, but the presence of low-level radiocarbon activity in organic matter immediately above the till (sample UW-85) suggests that the maximum advance of the Kobuk Glaciation might be of post-Illinoian age. Samples GXO523 and GXO522, from areas near Sinyalak Creek and the Helpmejack Lakes, merely represent the oldest organic matter recovered from trenches cut into permanently frozen sediments. Older organic-rich sediments undoubtedly are present within the permafrost at levels that could not be reached in the excavations.

North-Central Brooks Range

A sixfold glacial sequence (Table 4), established by U. S. Geological Survey mapping parties in the northern Brooks Range and adjacent foothills

(1) Trench cut into lacustrine deposits east of inlet stream 0.5 mile north of largest Help-mejack lake (Fig. 4). Bank 20 feet high, obscured at base by refrozen colluvium.

Unit	Thickness (inches)	Lithology
A	3	Surface organic mat.
B	12	Gleyed mineral soil; silt to fine sand. All depositional features destroyed by root activity.
C	0.5	Fragmented plant debris. Resembles detrital organic litter that accumulates along margins of open water bodies.
D	12	Silty lacustrine sediment; some shell fragments, plant debris, and fine-sand lenses present.
E	6	Charred organic horizon; wood fragments present.
F	12	Stratified peat.
G	4+	Marly lacustrine sediment; silty clay with shell fragments and fine plant debris. Radiocarbon date of 4765 ± 140 years B.P. (GX0522) from wood near upper contact. Extends below base of trench.

(2) Cutbank along Alatna River 1 mile upstream from bedrock gorge and range-front moraine belt (Fig. 5). Stands 34 feet high.

Unit	Thickness (feet)	Lithology
A	5	Silt, sand, and peat; in thin horizontal layers. Resembles over-bank deposits that cap modern flood plain.
B	11	Fluvial sand; cross-bedded. Contains current ripples and cut-and-fill structures. Grades laterally into horizontal layers of sand and organic detritus.
C	12+	Fine sand, silt, and clay; lacustrine. Wood fragments near upper contact are 6760 ± 90 years old (UW-84). Base of unit covered by colluvial apron that rises 6 feet above river level.

(3) Cutbank along Iniakuk Lake outlet stream 0.75 mile upvalley from moraine belt along range front (Fig. 5). Averages 30 feet high.

Unit	Thickness (inches)	Lithology
A	7	Surface organic mat.
B	15	Coarse sand to fine pebbles; stratified. Capped by 2 inches of eolian sand. Modified by podzolic soil profile development.
C	1	Clay band; Mn concentration at base.
D	120	Sand and gravel; cross-bedded, with cut-and-fill structures present. Medium sand dominant, but some beds consist of granules and small pebbles.
E	17–27	Medium sand; very well sorted; faint bedding. Upper contact shows erosional channel underlain by thin zone of oxide accumulation.
F	1–4+	Clay band. Permafrost table near upper contact. Contains small detrital wood fragments dated 31,000 ± 800 years B.P. (W-1427).
G	?	Medium sand, as in unit E. Almost entirely obscured by re-frozen colluvium.

(Detterman and others, 1958), has served as the basic framework for most subsequent Pleistocene studies in this region. The earliest glaciations, Anaktuvuk and Sagavanirktok, were marked by extensions of glaciers to positions 40 and 30 miles, respectively, north of the Brooks Range in their type areas. These advances were assigned pre-Wisconsin ages because of the modified appearance of their deposits. Anaktuvuk drift lacks identifiable moraines and is marked only by weathered erratics of resistant rock types and gently rounded hills composed of drift. The younger Sagavanirktok drift is marked by recognizable knob-and-kettle topography, and lateral and end moraines are present in some places. The Itkillik and Echooka Glaciations, of probable Wisconsin age, were less extensive and their features were little modified. The Alapah Mountain and Fan Mountain advances, confined to valley heads in higher parts of the range, subsequently have been assigned to neoglaciation (Porter, 1964; Porter and Denton, 1967).

Four substages of the Itkillik Glaciation were recognized by Porter (1964, 1967) in Anaktuvuk Valley. The Banded Mountain moraine, representing the oldest drift of Itkillik age (Table 4), outlines an ice lobe that terminated 23 miles north of the Brooks Range. The later Antler Valley moraine, broader and less subdued, is marked by numerous lakes and ponds and is largely composed of ice-contact gravel. Stratigraphic evidence indicated an intermediate episode, the Anayaknaurak readvance, which is poorly exposed topographically and probably occupied a relatively brief interval of time. The youngest drift sheet, Anivik Lake, is comparable to the Echooka drift of earlier workers but appears to represent a readvance of the same glacier that built the Antler Valley moraine. The Anivik Lake therefore is considered a substage of the Itkillik Glaciation. Radiocarbon dates from the Anaktuvuk region indicate that the Anayaknaurak readvance occurred shortly after 13,270 ± 160 years B.P. and that the wasting Anivik Lake glacier still occupied Anaktuvuk Pass about 7000 years ago. The Itkillik Glaciation probably encompassed all of late Wisconsin time; its four stades appear to correspond in age to subdivisions of the classical Wisconsin in central North America (Porter, 1964).

The Itkillik glacial sequence in the lower Alatna Valley is similar to the Itkillik succession in the Anaktuvuk region (Table 4). The outermost moraines, near Siruk Creek, form a series of subdued ridges that have been appreciably modified by postglacial mass wasting. The piedmont ice lobe associated with these moraines is comparable to the Banded Mountain lobe in size and in extent beyond the range front. The Chebanika moraine represents a relatively short-lived glacial event that may be comparable to the Anayaknaurak readvance in the Anaktuvuk area. The moraine is appreciably less modified than the earlier Siruk Creek drift, indicating that a significant period elapsed between these two events. A comparably long interval between the Anayaknaurak and Banded Mountain events is implied by Porter's correlations with the classical Wisconsin glacial sequence. Drift in the Helpmejack area resembles the Antler Valley moraine in its less subdued appearance, presence of numerous lakes and ponds, and abundance of ice-contact stratified drift. Moraines along the range front are similar to Anivik Lake (Echooka) deposits in their unmodified appearance and relative nearness to ice sources.

Kobuk Valley

Moraines around the southern end of Walker Lake near the head of the Kobuk Valley are identical in position and appearance to moraines along the Alatna at the range front (Pl. 5). The Walker Lake moraines are sharply defined and well-preserved; mass wasting has been negligible, and the steep flanks of moraine ridges contrast with the more subdued ridges of older drift farther downvalley. The Walker Lake Glaciation, as defined by Fernald (1964) was the last major glacial event in the Kobuk Valley. Subsequent glacial advances were restricted to upper parts of cirque-headed tributaries within the Brooks Range (Table 4).

Three older glaciations in the Kobuk region are reported by Fernald. The Ambler Glaciation, youngest of these events, is marked by lobate moraines along major valleys tributary to the Kobuk. Ice extended progressively farther south in eastern parts of the region and, near the head of the Kobuk Valley, approached the latitude of the Itkillik ice limit (for example, Norutak Lake moraines in Pl. 5). Moraine surfaces are marked by rounded knolls and undrained basins; their appearance is similar to that of the three oldest Itkillik moraines. Although Fernald fails to mention multiple moraine systems comparable to the Siruk Creek–Chebanika–Helpmejack drift sequence of the lower Alatna Valley, these apparently are present in the central Kobuk region. My current studies there, begun in 1968, indicate two or three glacial episodes of probable stadial rank separated by at least one major interstadial interval.

The Ambler Glaciation is separated from the older and more extensive Kobuk Glaciation by a major time break. Deposits assigned to the two glaciations show pronounced contrasts in surficial modification, and extensive eolian and alluvial sediments through much of the central Kobuk Valley probably were deposited during the interval between the ice advances. Organic matter near the base of these sediments is more than 33,000 years old (Fernald, 1964, p. 11–12). During the Kobuk Glaciation, ice coalesced in the Ambler Lowland and the Kobuk River Valley. Perhaps it formed an extensive piedmont lobe comparable in size to the Alatna Valley glacier, but the maximum extent of ice in the Kobuk region has not yet been determined.

An earlier ice advance filled the Kobuk Valley, extending west to Kotzebue Sound and south into the Koyukuk Lowlands where it probably was continuous with the oldest glaciation of the Alatna-Koyukuk region. Deposits related to this event have been assigned an Illinoian age (McCulloch and others, 1965). Relation of the earliest ice advance to the subsequent Kobuk Glaciation is unclear, as is the case in the Alatna region. Fernald states (1964, p. 9) that the Kobuk Glaciation "appears to be a separate and later glacial episode" but that "detailed mapping may indicate it to be a late phase of the earliest glaciation."

CONCLUSIONS

The Alatna Valley has been subjected to three major glaciations of Illinoian and Wisconsin age. Each glaciation is marked by a separate set of glacial facets along the flanks of the upper valley and by a distinct drift sheet farther south.

Drift limits of the oldest glaciation extend along the southern margin of the Koyukuk Lowlands and northwest into highlands along the southern flank of the Kobuk Valley. This glacial episode is correlated with an Illinoian ice advance that, farther west, filled the Kobuk and Noatak valleys and extended into Kotzebue Sound.

During the subsequent Kobuk Glaciation, ice advanced to a maximum position beyond the mouth of Alatna Valley; a later readvance or stillstand is marked by a moraine belt closer to the Brooks Range. Although bluff exposures along the Alatna, Koyukuk, and Kobuk Rivers aid interpretation of glacial events and ice limits, radiocarbon analyses of the sediments have provided neither precise dates nor closely limiting age estimates for the time of glaciation. The Kobuk ice advance may have occurred during early Wisconsin time, but a late Illinoian age also is possible.

Itkillik Glaciation is represented in the Alatna Valley by moraine belts near Siruk Creek, Chebanika Creek, the Helpmejack Lakes, and the southern flank of the Brooks Range. The moraines are believed to represent four successive stades of the Itkillik, and an additional stade possibly is represented by drift between the Helpmejack Lakes and the southern flank of the Brooks Range. All stades of Itkillik Glaciation are assigned to the late Wisconsin on the bases of slight postglacial modification of the drift, correlations with dated Itkillik events of the Anaktuvuk area, and radiocarbon dates from the Alatna Valley.

REFERENCES CITED

Chapman, R. M., Detterman, R. L., and Mangus, M. D., 1964, Geology of the Killik-Etivuluk rivers region, Alaska: U.S. Geol. Survey Prof. Paper 303-F, p. 325–407.

Coulter, H. W., and others, 1965, Map showing extent of glaciations in Alaska: U.S. Geol. Survey Misc. Geol. Inv. Map I-415.

Detterman, R. L., 1953, Sagavanirktok-Anaktuvuk region, northern Alaska, p. 11–12 *in* Péwé, T. L., and others, Multiple glaciation in Alaska: U.S. Geol. Survey Circ. 289, 13 p.

Detterman, R. L., Bickel, R. S., and Gryc, George, 1963, Geology of the Chandler River region, Alaska: U.S. Geol. Survey Prof. Paper 303-E, p. 223–324.

Detterman, R. L., Bowsher, A. L., and Dutro, J. T., Jr., 1958, Glaciation on the arctic slope of the Brooks Range, northern Alaska: Arctic, v. 11, p. 43–61.

Eakin, H. M., 1916, The Yukon-Koyukuk region, Alaska: U.S. Geol. Survey Bull. 631, 88 p.

Fernald, A. T., 1964, Surficial geology of the central Kobuk River Valley, northwestern Alaska: U.S. Geol. Survey Bull. 1181-K, 31 p.

Hartshorn, J. H., 1958, Flowtill in southeastern Massachusetts: Geol. Soc. America Bull., v. 69, p. 477–482.

Holmes, G. W., and Lewis, C. R., 1965, Quaternary geology of the Mount Chamberlin area, Brooks Range, Alaska: U.S. Geol. Survey Bull. 1201-B, 32 p.

Hopkins, D. M., 1963, Geology of the Imuruk Lake area, Seward Peninsula, Alaska: U.S. Geol. Survey Bull. 1141-C, 101 p.

—— 1967, Quaternary marine transgressions in Alaska, p. 47–90 *in* Hopkins, D. M., *Editor*, The Bering land bridge: Stanford, California, Stanford Univ. Press, 495 p.

McCulloch, D. S., 1967, Quaternary geology of the Alaskan shore of Chukchi Sea, p. 91–120 *in* Hopkins, D. M., *Editor*, The Bering land bridge: Stanford, California, Stanford Univ. Press, 495 p.

McCulloch, D. S., and Hopkins, D. M., 1966, Evidence for a warm interval 10,000 to 8,300 years ago in northwestern Alaska: Geol. Soc. America Bull., v. 77, p. 1089–1108.

McCulloch, D. S., Taylor, D. W., and Rubin, Meyer, 1965, Stratigraphy, non-marine mollusks, and radiometric dates from Quaternary deposits in the Kotzebue Sound area, western Alaska: Jour. Geology, v. 73, p. 442–453.

Peterson, J. A., 1965, Ice-push ramparts in the George River basin, Labrador-Ungava: Arctic, v. 18, p. 189–193.

Péwé, T. L., Hopkins, D. M., and Giddings, J. L., Jr., 1965, The Quaternary geology and archaeology of Alaska, p. 355–374 *in* Wright, H. E., Jr., and Frey, D. G., *Editors*, The Quaternary of the United States: Princeton, New Jersey, Princeton Univ. Press, 922 p.

Porter, S. C., 1964, Late Pleistocene glacial chronology of north-central Brooks Range, Alaska: Am. Jour. Sci., v. 262, p. 446–460.

—— 1967, Pleistocene geology of Anaktuvuk Pass, central Brooks Range, Alaska: Arctic Inst. North America Tech. Paper no. 18, 100 p.

Porter, S. C., and Denton, G. H., 1967, Chronology of neoglaciation in the North American Cordillera: Am. Jour. Sci., v. 265, p. 177–210.

Sainsbury, C. L., 1967, Quaternary geology of western Seward Peninsula, Alaska, p. 121–143 *in* Hopkins, D. M., *Editor*, The Bering land bridge: Stanford, California, Stanford Univ. Press, 495 p.

Schrader, F. C., 1904, A reconnaissance in northern Alaska: U.S. Geol. Survey Prof. Paper 20, 139 p.

GEOLOGICAL SOCIETY OF AMERICA, INC.

SPECIAL PAPER 123

Plant Fossils from a
Cary–Port Huron Interstade Deposit
and Their Paleoecological Interpretation

NORTON G. MILLER

Michigan State University, East Lansing, Michigan

WILLIAM S. BENNINGHOFF

University of Michigan, Ann Arbor, Michigan

ABSTRACT

Study of plant remains found intercalated between two till bodies in Cheboygan County at the northern tip of the Southern Peninsula of Michigan suggests that an open vegetation, floristically similar to a dwarf-shrub tundra or tundra communities within the tundra-boreal forest ecotone was present in this area sometime between 12,500 and 13,300 years B.P. The organic layer that has been correlated with the Cary–Port Huron (Lake Arkona) interstade consists mostly of bryophytes. Eight species of mosses; two kinds of leafy liverworts; achenes from *Carex tenuiflora* and/or *C. trisperma*; a perigynium of *Carex supina*; leaves of the arctic-alpine plants *Dryas integrifolia*, *Salix herbacea*, and *Vaccinium uliginosum* var. *alpinum*; and *Salix* twigs have been identified from the deposit. Pollen analysis of the bryophyte bed and associated sediments yielded spectra dominated by nonarboreal pollen, but a significant amount of spruce pollen is also present. Evidence is given that indicates that some of the pollen is rebedded. The macrofossils suggest the presence of distinct communities on wet and dry sites.

CONTENTS

INTRODUCTION

The discovery during the summer of 1965 of a thin layer of bryophytes beneath several meters of pink till in the northern part of the Southern Peninsula of Michigan promised to provide useful information on the Pleistocene paleoecology and chronology of the region. The geological setting and significance of the site have been described (Farrand and others, this volume), but it will be useful here to review this information, as it is essential background for a discussion of the plant fossils in the deposit, the purpose of the present paper.

The moss horizon was uncovered on the farm of Ed Ginop 0.5 mile north of Munro Lake in Cheboygan County (central Sec. 2, T.37N., R.3W.) during construction of a farm pond. Samples from the moss bed were brought to the attention of Robert Zahner at the nearby University of Michigan Biological Station, who undertook to collect more samples and to examine the site in detail that summer. From the data collected at that time and from core drilling operations carried out in late 1967, the stratigraphy of the site was determined to consist of nearly 1 m of stratified Lake Algonquin sand and gravel at the top and beneath this, in order, about 3 m of reddish-brown, sandy, clay till, about 4 cm of reddish-brown indistinctly laminated lacustrine clay and silt, 1 cm of compressed bryophyte material, approximately 70 cm of sorted sand and gravel with finely comminuted organic debris incorporated into the upper 2 to 3 cm, and at least 7 m of another reddish-brown, sandy, clay till identical in color and texture to the till above the organic layer. By augering and drilling, the areal extent of the moss bed was found to be about 400 m^2.

The upper part of this stratigraphic sequence suggests that the bryophyte bed was covered by lacustrine silts and clays deposited in a proglacial lake ponded by an advancing ice sheet which later overrode the site and left the upper till. Since previous work in this area of Michigan had demonstrated the presence of red Valders till (Spurr and Zumberge, 1956), it was initially felt that the organic horizon might be equivalent to the Two Creeks forest bed of eastern Wisconsin where Valders till and lacustrine sand overlie about 10 cm of organic material. But whereas wood from the Two Creeks forest has recently been dated at 11,850 ± 100 years B.P., an average based on six different samples (Broecker and Farrand, 1963), the age of the Ginop farm site moss bed is about 1000 years older, somewhere between 12,500 and 13,300 years B.P. The absence of either a Two Creeks organic horizon or a weathered zone to represent this time suggests that the Two Creeks Interstade and perhaps the Valders Stade as well are not represented by deposits at the Cheboygan County site.

The moss bed can, however, be correlated with the Cary–Port Huron (Lake Arkona) interstade. Evidence for this event had previously been drawn from the occurrence of the low Lake Arkona stage between the preceding Lake Maumee and the following Lake Whittlesey higher lake stages in the Erie and Huron basins. These stages suggest withdrawal of ice during Lake Arkona time and a subsequent readvance that has been correlated with the deposition of the Port Huron moraine during Lake Whittlesey time. Published dates for these events include: 13,600 ± 500 years B.P. for wood between the deposits of Lake Arkona and Lake Whittlesey at Cleveland, Ohio (W-33; Suess, 1954), 12,920 ± 400 years B.P. for wood from a peaty zone beneath Lake Whittlesey beach gravels at Parkertown, Ohio (W-430; Rubin and Alexander, 1958), and 12,800 ± 250 years B.P. for wood fragments imbedded in beach sediments of Lake Whittlesey near Bellevue, Ohio (Y-240; Barendsen and others, 1957). Accepting this correlation, Farrand and others (this volume) postulated the following sequence of events to explain the stratigraphy at the site:

(1) Deposition by Cary ice of the lower till and retreat of this ice to some point north of the site;

(2) Growth of the bryophyte bed on outwash sand;

(3) Flooding of the bryophyte bed by melt water from advancing Port Huron ice and deposition of lacustrine silts and clays over it, and, finally, deposition of the upper red till by Port Huron ice;

(4) Withdrawal of Port Huron ice during Two Creeks time to a point just north of the Petoskey-Cheboygan lowland that served as a drainage channel between the Lake Michigan and Lake Huron basins, site of bryophyte bed remaining covered by ice;

(5) Advance of Valders ice southward in Michigan;

(6) Withdrawal of the ice from Port Huron and Valders advances and the synchronous expansion of Glacial Lake Algonquin.

The plant fossils in the bryophyte bed in general support this interpretation. They consist mainly of compressed but otherwise well-preserved mosses with a few liverworts, some carbonized leaves, seeds, and twigs, and, in addition, pollen. The pollen spectrum from the bryophyte bed suggests an open vegetation dominated by members of the Cyperaceae and other non-arboreal plants with spruce trees probably growing some distance from the site of pollen deposition. The mosses and other macrofossils represent species that are consistent with this interpretation. Possible modern analogues of this vegetation will be discussed later.

The usefulness of supplementing pollen spectra with data from the study of macrofossils to effect a better understanding of the vegetation and vegetation changes during late Wisconsin time has recently been emphasized by a number of workers (Argus and Davis, 1962; Baker, 1965; Schofield and Robinson, 1960; Watts and Winter, 1966). Late-glacial sites have received particular attention in this regard with the result that leaves of such tundra plants as *Dryas integrifolia* and *Salix herbacea* have been recognized in several deposits. The occurrence of these fossils has supported the belief that tundra or a patchwork of boreal forest and tundra existed near the glacier margin during the waning stages of glaciation. Since it is unlikely that such fragile structures as leaves and many types of seeds would be transported for long distances by moving water and still remain identifiable, their occurrence in the fossil condition is assurance that the plants that produced them lived at or near the site of deposition. This is in contrast to the pollen record that includes a representation of the pollen produced not only by the local vegetation at the sampling point but by the extra-local vegetation adjacent to the sampling point and the regional vegetation on the upland some distance from the sampling point as well, as has been demonstrated for Minnesota by Janssen (1966) and for Cheboygan County by Benninghoff (1960).

Certain plants that grow great distances from the sampling site are also occasionally represented in pollen spectra. The portion of a pollen diagram that reflects changes in the regional pollen rain is best used to interpret major changes in vegetation and climate.

Examples of combined macro- and microfossil studies of plant fossils in interstadial and interglacial deposits are less common. The thorough examination of the Two Creeks forest bed at a number of sites, however, is a noteworthy exception (*see* review by Black and others, 1965). Cones of both black and white spruce; needles, twigs, and wood of balsam fir; larch cones and wood; *Populus* wood; a bracket fungus; 23 different mosses, 19 of which were identified to the species level, 3 to the generic level, and 1 to the varietal level using currently accepted names (Crum and others, 1965); shells of 7 species of

Mollusca; borings of bark beetles; and 2 genera of testate protozoa have been found either in the Two Creeks forest bed itself or in the sediments directly above or below this deposit. Many of these finds complement the pollen spectra. At the type Two Creeks location along the western shore of Lake Michigan near the base of the Door Peninsula in Wisconsin, spruce pollen dominated all analyzed levels of forest bed except the bottom level in which the pollen of the shrub *Shepherdia canadensis* reached more than 95 percent of the total pollen (West, 1961). Small percentages of additional trees, shrubs, and herbs are also represented in these spectra. Pollen analysis of another Two Creeks site near Green Bay, Wisconsin, 30 miles northwest of the previously mentioned locality, has demonstrated a sequence in which sedge pollen, dominant (60 to 80 percent) in the inorganic sediments beneath but including the basal part of the forest bed, is replaced by spruce pollen that reaches a maximum near the top of the forest bed (Schweger *in* Black and others, 1965). This change presumably documents invasion and establishment during Two Creeks time of a spruce forest on surfaces where sedges were previously dominant.

Terasmae's (1960) study of pollen in the Pleistocene interglacial beds near Toronto, Ontario, has similarly added important information to that previously derived from the leaf and fruit fossils in the deposit.

ACKNOWLEDGMENTS

We gratefully acknowledge assistance from Dr. F. J. Hermann, Forest Service Herbarium, U.S. Department of Agriculture, for identification of vascular plant macrofossils belonging to the Cyperaceae; Dr. Frank Eggleton, University of Michigan, for identification of molluscan material; Dr. Howard Crum, University of Michigan Herbarium, for critical review of bryophyte identifications; and Dr. John H. Beaman for use of the facilities of the Beal-Darlington Herbarium of Michigan State University. Dr. Robert Zahner, University of Michigan, provided opportunity for us to work on the collected materials. He and Dr. William R. Farrand, also of the University of Michigan, furnished helpful criticism throughout the study.

METHODS

In the effort to observe the stratigraphy at the site better, a small pit located at the southeastern corner of the farm pond was dug with a backhoe during the summer of 1965. Chunks of sediment containing an undisturbed lacustrine clay–bryophyte bed–sand sequence were removed from the deposit during the operation. Samples for pollen analysis were obtained from these sediment blocks by N. G. Miller. The method involved splitting the block vertically to produce a clean, uncontaminated face from which samples of the three sediment types were removed with a clean knife blade. Samples were stored in labelled, stoppered vials until they were prepared for analysis in the laboratory. Several series of samples for pollen analysis were obtained from different blocks, and one set was analyzed by each of us. A large amount of the compressed bryophyte bed was also collected at this time for macroscopic study.

Samples for pollen analysis were prepared in the usual manner involving successive treatments of 10 percent HCl, 52 percent HF, and 10 percent KOH and acetolysis (Faegri and Iversen, 1964). The inorganic sediments were strongly calcareous. Samples prepared by Miller, which contained a large amount of heavy minerals, especially pyrite, were put through heavy-liquid separation using zinc chloride solution (specific gravity 1.93). Counts were made using equispaced traverses controlled by a calibrated mechanical stage across the area of a cover slip. Counting was done routinely at a magnification of 250 diameters. Critical grains were examined under oil immersion.

Samples of the bryophyte bed collected for macroscopic analysis were soaked in distilled water and gently teased apart. Clays, silts, and fine organic matter were removed by washing them with distilled water through a 250-μ sieve into a beaker. The fine sediments were devoid of recognizable fossils. Nearly 200 ml of organic material were retained by the sieve and were available for study. This material was examined under a dissecting microscope against a white background. Potentially identifiable plant remains and a representative suite of the mosses were removed from the sample with a camel's hair brush or a large bore pipette. The fossils were refrigerated in vials containing only distilled water. No fungal activity was noted.

Identifications were checked against herbarium specimens or, in the case of most of the mosses, the determination was verified or made by Dr. H. Crum of the University of Michigan. The photographs were taken with a 35-mm camera equipped with a bellows and a portrait lens.

THE MACROFOSSILS

Mosses make up nearly 100 percent of the organic material in the bryophyte bed, although a small amount of several species of liverworts and leaves, seeds, and twigs of a number of vascular plants has been found. The mosses are well-preserved and features necessary for identification are readily observed. They are less strongly blackened than the vascular plant remains and are similar in color to modern collections of the species. Leaves of vascular plants isolated from the moss bed have in general kept their three-dimensional form and, although leaf fragments are present, whole leaves showing little damage are frequent. Identifiable macrofossils are restricted to the moss bed. Small bits of plant debris have also been observed in the lacustrine deposit above the moss bed and in the upper part of the sand beneath it. Judging from the shape and arrangement of the cells in the debris, most of it is derived from either mosses or graminoid plants.

Mosses

Bryum cryophilum Mårt. Several stems bearing the concave, rounded leaves characteristic of this species were found. In northern Scandinavia the plants grow in low cushions, generally on alluvial sand or silt beside streams or in areas marked by water seepage (Mårtensson, 1956). Dr. Crum, who has observed the species in northern North America, has reported to us that it is found in exposed, moist, sandy or gravelly habitats on this continent as well. Steere (1947) considers it a truly arctic species. A closely related lowland plant, *B. tortifolium*, has been reported from the Two Creeks forest bed (as *B. cyclophyllum*; Cheney, 1931).

Range: circumpolar; arctic North America and Greenland.

B. pseudotriquetrum (Hedw.) Gaertn., Meyer, and Scherb. This moss, which is apparently quite rare in the deposit, has lanceolate, acuminate leaves that are long-decurrent down the stem. It is a common species today in the Straits of Mackinac region of Michigan where it occupies a variety of habitats, usually beneath forest cover but often at exposed sites as well. It grows on wet soil or humus and occasionally on inorganic substrates in swamps or near streams and lakes (Crum, 1964).

Range: circumboreal; found nearly throughout the United States but mostly in boreal and arctic regions.

Campylium stellatum (Hedw.) C. Jens. Although only moderately abundant in the deposit, this species is at the present time an important member of a moss community that is commonly found at marshy, calcareous sites, particularly along the shores of Lake Michigan and Lake Huron in the Straits of Mackinac region. Other noteworthy members of this community are *Calliergon trifarium, Cinclidium stygium, Drepanocladus vernicosus, D. revolvens* var. *intermedius,* and *Scorpidium scorpioides.* All of these mosses are strongly calciphilous; some grow directly on marl.

Range: circumboreal; across the continent in the northern United States and Canada, south to Pennsylvania, Ohio, and Colorado.

Catoscopium nigritum (Hedw.) Brid. This is one of the two most abundant mosses in the bryophyte bed. Fairly large clumps composed of many individual plants matted together by the dense basal development of rhizoids characteristic of this species were found, suggesting the absence of excessive abrasion in moving water during the formation of the bryophyte bed. Although it can be found today forming large cushions directly on moist stabilized beach sand well above the zone of disturbance by wave action, it is rare in the Straits of Mackinac region of Michigan, being restricted to stations along the calcareous strand of Lake Huron and Lake Michigan (Crum, 1964).

Range: circumboreal; across North America, south to New York state, the Great Lakes, Iowa, and Montana.

Ditrichum flexicaule (Schwaegr.) Hampe. Only a few stems of this moss were found in the sample of the bryophyte bed that was examined. It also presently grows near the fossil site and, like the preceding species, is often found on sand along the shores of the northern parts of Lake Michigan and Lake Huron but in drier situations. In the Hudson Bay lowlands it grows at wetter sites such as moist forests and rich fens (Persson and Sjörs, 1960). It is a decided calciphile that Steere (1947) considers one of the most common of all alpine, arctic, and subarctic Bryophyta.

Range: circumboreal; across northern North America south to the northernmost United States.

Drepanocladus aduncus var. *polycarpus* (Bland. *ex* Voit) Roth. Where this species grows today, it is often very abundant, forming submerged or emergent loosely woven mats in shallow, frequently calcareous, water. It is only a minor component of the fossil moss assemblage, however. Crum (1964) lists twelve stations in the four counties bordering the Straits of Mackinac and notes that it grows on wet soil in swamps, rich fens, meadows, and swampy woods.

Range: circumboreal; widespread in temperate and boreal North America.

D. revolvens var. *intermedius* (Lindb. *ex* C. J. Hartm.) Rich. and Wallace. This moss, which was identified by Crum in the sample of the bryophyte bed given to him for study, is also a member of the present flora of the region around the fossil site. It is a plant of wet soils in swamps, rich fens, and meadows. The variety typically grows in calcareous habitats.

Range: circumboreal; northern United States and Canada, across the continent.

Scorpidium turgescens (T. Jens.) Loeske. This is the most abundant fossil in the moss bed. In the Owen Sound region of Ontario the species has been found growing emergent in pure stands 100 feet across and reaching a maximum length of about 20 cm over sand in shallow calcareous marshes (Conard, 1938). It is, however, a very rare plant, especially along the southern edge of its range. The species is known from only four stations in Michigan: Bois Blanc Island at the eastern entrance to the Straits of Mackinac (Darlington, 1964), Drummond Island at the eastern extremity of the upper peninsula of Michigan (Fork Kamp, *Koch 2616*, MICH), near Charlevoix along the shore of Lake Michigan (*Mazzer s.n.*, UMBS), and on Summer Island off the tip of the Garden Peninsula just west of Manistique (*Halbert 1151*, MICH). The species is a calciphile that is characteristic of rich fens and limy sloughs. It is one of several species whose presence in Great Britain during much of Wisconsin time has been documented by fossils (Dickson, 1967). Although the only confirmed station in Great Britain at the present time is in central Scotland (Ben Lawers), a late Wisconsin occurrence in southern Scotland and early and middle Wisconsin occurrences in the southern lowlands of England suggest an extinction northward during these times.

Range: circumboreal; northernmost North America, south to the Great Lakes.

Liverworts

Fragmentary remains of several species of leafy Hepaticae were found sparsely represented in the moss bed. Poor preservation of the leafy stems and the absence of perianths and male and female inflorescences make identification to species impossible, but at least two genera are present. One of these, clearly a species of *Cephalozia*, has more or less round, bilobed leaves with large thin-walled cells. The outer cortical cells of the stem are also large and thin-walled. The other hepatic may be a member of the Lophoziaceae. Its leaves have two acute lobes and the leaf cells are smaller and thicker walled and have trigonous enlargements at the corners. Leaf insertion is oblique. Certain species of *Cephalozia* and a member of the Lophoziaceae are not unexpected associates of the mosses mentioned above.

Vascular Plants

Carex spp. Achenes of one or perhaps several *Carex* species were abundantly represented in the bryophyte bed and in the upper part of the sand directly beneath. One well-preserved perigynium was also isolated from the bryophyte layer. Photographs of some of these fossils were sent to Dr. F. J. Hermann, who made the identifications upon which the following discussion is based.

The perigynium can be referred to *Carex supina* Willd., a circumpolar species that grows on rock ledges and in dry, exposed, sandy or gravelly areas (Polunin, 1959; Pl. 1, figs. 2, 3). Porsild (1957) reports that it is found at noncalcareous sites. Since nearly all of the other plants found in the deposit are calciphiles, *Carex supina* may have been a member of the "upland" flora where it perhaps occupied sandy slopes above the basin that were leached free of carbonates. *Carex supina* is an aggregate species composed of three subspecies of which only subsp. *spaniocarpa* (Steud.) Hult. is found in North America (Hultén, 1964). It is distributed across the arctic and subarctic from central Alaska east to Baffin Island and south to the Nelson River in northern Manitoba. Outlying populations occur in northeastern Minnesota and southern Manitoba.

The achenes can be identified with less assurance (*see* Pl. 1, fig. 1). They closely match certain species in the section Dispermae, particularly *Carex tenuiflora* Wahlent. and *C. trisperma* Dewey. The former is another circumpolar species that in North America is mainly subarctic in distribution with southward extensions to northern New Jersey, New York, and southern

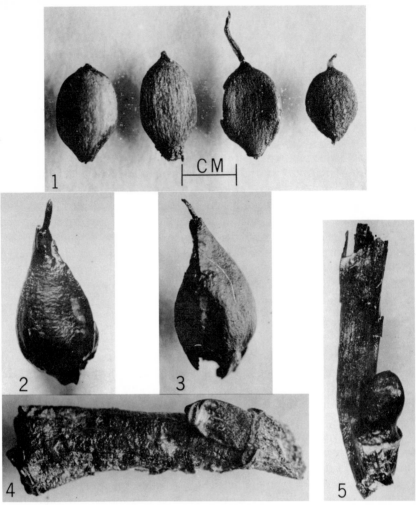

Figure 1. Achenes of *Carex tenuiflora* Wahlent. and/or *C. trisperma* Dewey (fossil).

Figure 2. Perigynium of *Carex supina* Willd., front view (fossil).

Figure 3. Same perigynium as in figure 2, side view (fossil).

Figure 4. *Salix* twig (fossil).

Figure 5. *Salix* twig (fossil).

FOSSIL ACHENES AND PERIGYNIUM OF *Carex* spp. AND FOSSIL TWIGS
OF *Salix* sp. or spp.

Scale for figures 2 to 5 same as in figure 1.

Michigan (Hultén, 1962). *Carex trisperma* occurs from southern Labrador to Saskatchewan south to Minnesota, northern Indiana, and Maryland and down the Appalachians to the Great Smoky Mountains (Fernald, 1950). Both are species of bogs, mossy forests, and pond margins and other open areas.

Dryas integrifolia M. Vahl. Of the fossils of flowering plants represented in the bryophyte bed, the blackened leaves of this plant occur in greatest abundance. The size range of the fossil leaves is somewhat smaller than that present in many herbarium specimens of the species, but this may be the result of selective transportation of smaller leaves to the moss bed. The largest leaf found was 8 mm long and 1 mm wide; the smallest leaves measured 1.75 to 2.00 mm by 0.80 mm. Most fell about midway between these extremes, however. Apart from the size difference, the fossil leaves are identical in shape and venation to modern specimens. A comparison of fossil and modern leaves of the same size demonstrates their remarkable similarity (Pl. 2, fig. 1, modern leaves from western Greenland, Nugsuak Peninsula, *Cornell Party*, Peary Voyage, 12 August 1896, MSC). The dense tomentum on the underside of the leaves was not observed in any of the fossil specimens.

According to Polunin (1959), *Dryas integrifolia* is "one of the most ubiquitous and ecologically important plants of the New World Arctic, dominating many dry and exposed clay or gravel 'barrens' and open sandy heaths with its domed tussocks, but also occurring as a looser growth in sheltered and well-watered situations" (p. 265). Porsild (1957) also notes its common occurrence in the arctic and calls attention to its role as a pioneer in calcareous habitats. Lawrence and others (1967) have recently demonstrated the very significant role of *Dryas* as a nitrogen-fixing species (through root nodule bacteria) among pioneers on fresh glacial deposits. It is also found in alpine areas in western North America south to Montana (Pl. 3), where it is mostly represented by two subspecies, *D. integrifolia* subsp. *subintegrifolia* and subsp. *sylvatica* (Hultén, 1959). Only the typical subspecies occurs in eastern North America (Hultén, 1959). It prefers calcareous gravels and rocks and is found as far south as the Gaspé Peninsula in eastern Quebec (Fernald, 1950). Scattered stations occur northward in maritime Canada.

Leaf fossils of *Dryas integrifolia* have been reported from a number of late-glacial deposits in Minnesota (Cushing, 1967; Watts, 1967; *see* Pl. 3). The absence of macrofossils of boreal forest species and the occurrence of leaves and seeds of other arctic-alpine plants in these deposits suggests that a dwarf-shrub tundra was present at the sampling sites prior to 10,500 years B.P. in northeastern Minnesota and earlier at the more southern stations (Cushing, 1967; Watts, 1967). Leaves of *D. integrifolia* have also been found with fossils of other tundra plants in late-glacial sediments near Cambridge, Massachusetts (Argus and Davis, 1962).

Figure 1. Fossil and modern leaves of *Dryas integrifolia* M. Vahl; from left: upper surface (fossil), upper surface (modern), lower surface (fossil), lower surface (modern).

Figure 2. Leaves of *Salix herbacea* L.; at left, lower surface (fossil), at right, lower surface (modern).

Figure 3. Leaves of *Vaccinium uliginosum* var. *alpinum* Bigel.; at left, lower surface (fossil), at right, lower surface (modern).

FOSSIL AND MODERN LEAVES OF *Dryas integrifolia* M. Vahl, *Salix herbacea* L., AND *Vaccinium uliginosum* var. *alpinum* Bigel.

Range: Predominantly North American (*see* Pl. 3); eastern Greenland across arctic America to and including easternmost arctic Asia, south to Gaspé Peninsula and Montana, with isolated stations on the northern shore of Lake Superior and at Lake Mistassini, Quebec. Additional disjunct stations have recently been reported from Montana (Bamberg and Pemble, 1968) and southeastern New Brunswick (Roberts, 1965).

Salix herbacea L. Several whole leaves and a number of fragments were found that could be definitely referred to this species. The round, somewhat obovate shape of the leaf blade, its truncate apex, and the short, forward-pointing teeth along the margin are very characteristic (*see* Pl. 2, fig. 2, modern leaf from Mt. Washington, New Hampshire, *Allen s.n.*, MSC). The fossil leaf in the illustration was originally folded along the midvein and had to be broken and pieced together for photographing. A portion of the apex was lost at this time.

Salix herbacea is an arctic-alpine species whose occurrence in the northern part of its range reflects in part the presence of late snow patches (Porsild, 1957). It occupies a number of distinct habitat types including sunny, sheltered, well-drained, south-facing slopes, barren hill summits, dry upland tundra sites, and depressions in *Dryas* barrens (Polunin, 1948). In eastern North America it grows on the alpine summits of the highest mountains from the Adirondacks of New York northward.

Range: Amphi-Atlantic; eastern arctic and subarctic North America west to Great Bear Lake, south to Churchill in the midcontinent, and east to New York state.

Additional leaf types found in the deposit may be from other species of *Salix*. An attempt was made to identify these but was discontinued because of the polymorphic nature of the leaves of many dwarf willows. *Salix* twigs were also present in the moss bed. They have somewhat elongated, bullet-shaped buds covered by one scale. The buds are subtended by shallow U-shaped leaf scars (Pl. 1, figs. 4, 5). Some of the twigs are similar in size and organization to those of *Salix herbacea*, but other *Salix* species may be represented as well.

Vaccinium uliginosum var. *alpinum* Bigel. This species is well represented by leaf fossils in the deposit. They are oval to obovate in shape, have entire margins, and frequently show small acute apices. Reticulate venation is often apparent on the underside of larger leaves (*see* Pl. 2, fig. 3, modern leaf from near Churchill, Manitoba, *Gillis 3386*, MSC). The fossil leaf illustrated is smaller than most of the ones found but was photographed because of its flatness and because in larger specimens the lamina was usually folded back on itself. Although a conclusive identification based on leaves alone is perhaps unwise, the small size of the leaf fossils indicates that they can be referred to the var. *alpinum* Bigel. and are so treated here.

Ecologically *Vaccinium uliginosum* var. *alpinum* is a plant of acid soils at both dry and wet sites (Porsild, 1957) and grows not only in more favorable, sheltered situations but also in peaty marshes and open gravelly areas throughout the low- and middle-arctic region (Polunin, 1959). In eastern North America it is a frequent plant of upper exposed mountain slopes.

Range: circumpolar; across boreal and arctic North America, south to northern New York, Michigan, and Minnesota (Fernald, 1950).

Molluscs

The only animal macrofossils found in the deposit were very small snail shells in the bryophyte stratum. Dr. Frank Eggleton identified the collection, which consisted of three immature and two subembryonic snails and two fragments of snails, all of which are provisionally referred to *Stagnicola emarginata angulata*. This snail species has a wide range today in North America and is common in the upper Great Lakes region.

THE POLLEN SPECTRA

Seven samples were analyzed by each of us for pollen. These were collected in duplicate in two series so that we both had a stratigraphically identical set to study. The first series consisted of one sample from the bryophyte bed and two samples from the lacustrine deposit above it and was taken from a sediment block along a vertical line crossing the different lithologies. The second series, composed of four samples, was collected in a similar manner from a different sediment block and included a sample from the upper part of the sand beneath the organic bed in addition to three samples in the same stratigraphic positions as those of the first series. Both sediment blocks may originally have been adjacent. It was impossible to determine whether or not they were contiguous, however, because of the crudeness of the backhoe excavation and the presence of muddy, standing water that accumulated in the hole below a depth of 2 m and obscured the bottom. In any case, the sample series could not have been any farther apart than the 2-m by 4-m areal limit of the pit.

Our counts agreed closely even though they were made in different laboratories using somewhat dissimilar techniques and equipment. Although the pollen spectra presented in Figure 1 are from data collected by Miller, the spectra from the duplicate samples of this series counted by Benninghoff are essentially the same. A close agreement between pollen spectra from stratigraphically equivalent samples in the two sediment blocks was also noted. One difference was the greater abundance of pollen on slides made from residues that were treated with zinc chloride to remove the heavier inorganic sediments remaining after the samples had been treated with hydrochloric and hydrofluoric acids. This process, if done correctly, will not affect the numbers

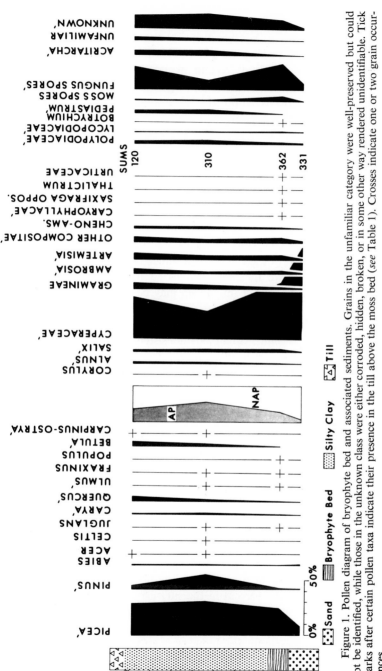

Figure 1. Pollen diagram of bryophyte bed and associated sediments. Grains in the unfamiliar category were well-preserved but could not be identified, while those in the unknown class were either corroded, hidden, broken, or in some other way rendered unidentifiable. Tick marks after certain pollen taxa indicate their presence in the till above the moss bed (see Table 1). Crosses indicate one or two grain occurrences.

or types of pollen originally present in the sediment; rather it concentrates the pollen for easier counting. Another disparity was the recognition by each of us of pollen types not noted by the other. This may reflect differences in our familiarity with the pollen of certain plants rather than an inherent dissimilarity of the pollen content of the samples.

Only four of the fourteen possible spectra are included in the pollen diagram. The others provide little additional information and are therefore omitted. Percentages of tree, shrub, and herb pollen were calculated using the sum of the numbers of pollen grains in the three groups as the percentage base. The sum was different for each spectrum. The percentages of items in the miscellaneous category were calculated by adding the sum of these in each spectrum to the basic percentage base of that spectrum and using this figure as a new percentage base.

The most striking feature of our diagram is the presence of a large amount of pollen from herbs and shrubs. Of the pollen of these nonarboreal plants the most abundant is that of the Cyperaceae, which varies from 38 percent in the lacustrine deposit above the moss bed to 85 percent in the sand directly beneath it. Although some of this pollen may be assigned to *Eriophorum* on the basis of its irregular shape and large size, judging from the fairly abundant macrofossil remains of *Carex*, a large amount of it was probably produced by members of this genus. Small but consistent amounts of *Alnus*, *Salix*, Gramineae, *Ambrosia*, *Artemisia*, other Compositae, and Cheno-amaranth pollen are present in most of the spectra. Single pollen grains of other herbaceous plants, including a member of the Caryophyllaceae, *Thalictrum*, and a member of the Urticaceae, were restricted to the moss bed. Pollen indistinguishable from reference slide material of *Saxifraga oppositifolia* was also found in this level. Three grains were encountered while counting to the basic sum, and additional grains were noted while scanning after the sum was reached.

Spruce pollen ranges from a low of 8 percent in the spectrum beneath the moss bed to a high of 31 percent in the silty clay above it. Wing-tip–to–wing-tip measurements of well-preserved spruce grains indicate that both black and white spruce are represented. One-third to two-fifths of measurable spruce pollen was less than 100 μ, the point usually used to separate the smaller black spruce grains from the larger grains of white spruce (West, 1961). *Pinus* pollen is weakly represented in the moss bed and in the sand beneath it (<1 percent) but reaches a high of 13 percent in the silty clay above. A small amount of *Abies* pollen occurs in three of the four spectra. *Betula* is found only in the two levels above the moss bed.

The pollen of temperate deciduous trees appears regularly in the spectra. *Carya* and *Quercus* are uniformly represented in the upper three levels, while pollen of *Acer*, *Celtis*, *Fraxinus*, *Juglans*, *Populus*, *Ulmus*, and *Carpinus-Ostrya* are present as sporadic occurrences of one or two grains. Such pollen is regularly found in low percentages in the inorganic sediments deposited during the spruce– or spruce-sedge–dominated vegetation of late glacial time. Although representatives of some of the above genera may possibly have become established in especially favorable locations of restricted extent near the depositing area, it is more likely that this pollen was either redeposited from older sediments or arrived at the point of deposition by long-distance transport. The occurrence of small amounts of *Carya, Fraxinus, Juglans, Quercus,* and *Ulmus* pollen in atmospheric samples collected in the subarctic at Churchill, Manitoba, about 400 miles beyond the range of these genera is evidence supporting the latter explanation (Ritchie and Litchi-Federovich, 1967).

We have no way of evaluating the effect of long-distance transport on the pollen spectra discussed here. The till directly above the moss bed was examined as a possible source of pollen available for rebedding. It perhaps would have been better to examine the lower till for pollen since this must have been at the surface when the moss bed was forming, but a sample of this was not available for study. However, most of the lacustrine sediments above the moss bed were probably carried to the site by glacial melt water after having been derived from the ice which later deposited the upper till.

The results of the analysis are given in Table 1. Pollen was extremely sparse in the till. No quantitative estimate of pollen frequency was made, but an initial sample of 20 gm of dry till produced a very small organic residue. Three 22- × 22-mm cover slips were completely traversed and a sum of only 77 recognizable grains was reached. It is interesting, however, that some deciduous tree pollen was found. Much of it is abraded or broken, as is pollen of the same type in the bryophyte bed and

TABLE 1. POLLEN ANALYSIS OF TILL ABOVE MOSS BED

	No. Grains	%
Trees		
Picea	17	22.1
Pinus	20	26.0
Carya	4	5.2
Quercus	2	2.6
Ulmus	2	2.6
Tilia	1	1.3
Carpinus-Ostrya	2	2.6
Betula	3	3.9
Shrubs		
Alnus	4	5.2
Salix	1	1.3
Herbs		
Cyperaceae	7	9.1
Ambrosia	4	5.2
Artemisia	4	5.2
Other Compositae	5	6.5
Caryophyllaceae	1	1.3
Percentage Base A	77	
Miscellaneous		
Polypodiaceae	6	2.1
Lycopodiaceae	2	0.7
Sphagnum	1	0.4
Bisaccate fragments	78	27.4
Fungus spores	1	0.4
Pediastrum	14	4.9
Acritarcha	55	19.4
Triporate corroded	10	3.5
Unknown	41	14.4
Trees, shrubs, herbs	77	27.0
Percentage Base B	285	

associated sediments, but perfectly preserved specimens were also present. This implies that the temperate deciduous pollen in our counts could have been largely redeposited. The presence of a fairly high percentage of *Pinus* pollen in the silty clay above the organic layer is likely also to be a result of rebedding.

On the other hand, the occurrence in the till of pollen that is logically expected to have been produced by the type of vegetation indicated by the macrofossils suggests that percentages of *Picea*, Cyperaceae, *Ambrosia*, *Artemisia*, and other herbaceous plants may be inflated in our spectra due to redeposition. We employed the usual technique of keeping track of identifiable parts of bisaccate grains and then dividing the sum of these by a suitable denominator chosen with reference to the size of the fragments, giving us a figure that was equal to the number of reconstructed whole grains. This technique may have further enlarged the percentage of *Picea* pollen in our spectra because of the high frequency of bisaccate fragments of this genus in the till.

The occurrence of Acritarcha belonging to Paleozoic members of the genera *Baltisphaeridium* and *Veryhachium* in both the till and our spectra is additional proof that rebedding has taken place. The presence of these palynomorphs is of no importance to the paleoecology of the site, since presumably they were derived from the early Paleozoic carbonate bedrock of the region.

It is apparent then that only the two lowest of our spectra that contain the smallest amount of silt provide percentages that are least influenced by the effects of redeposition of pollen exhumed from older or perhaps in part contemporaneous deposits. Fluctuations in percentages must be interpreted with this in mind.

DISCUSSION

Both the pollen diagram and the macrofossils, for the moment interpreted without reference to the redeposited grains, imply an open vegetation dominated by members of the Cyperaceae and other herbaceous plants. Spruce and perhaps additional trees were present but probably occurred scattered and at some distance from the bryophyte bed. The presence of *Dryas integrifolia* and other tundra heliophytes necessitates the absence of a closed forest around the site, and the relatively low percentages of spruce pollen are further evidence that this was the case. In three out of four spectra the ratio of nonarboreal pollen (NAP) to arboreal pollen (AP) is greater than 1, the standard usually used to characterize the occurrence of tundra in a pollen diagram (*see* Livingstone, 1955). Although this ratio must be used with caution, it nevertheless emphasizes the importance of nonarboreal pollen in the spectra. The pollen diagram and the macrofossils suggest the presence of several distinct tundra communities.

The effect of redeposited pollen on the pollen diagram must, however, be evaluated. We have hesitated to simply subtract pollen suspected of having been rebedded because of the possibility that not all was derived from the till. Some of it, especially the pollen of deciduous forest trees, most likely was redeposited from this source and therefore can be removed from the diagram. Pollen of spruce, Cyperaceae, and certain other nonarboreal taxa, however, although also present in the till, may have been redeposited penecontemporaneously as well (Cushing, 1964). This type of redeposition can take place whenever slope wash occurs, thus tending to overrepresent certain pollen types of local derivation. If slopes adjacent to the bryophyte bed were extensive and if the vegetation on them was sufficiently sparse to facilitate drainage into the basin, then the pollen contributed to the sediments would not be derived solely from airborne grains but would include pollen washed in from the surrounding area. The fact that slope wash or some other type of mixing did take place is indicated by the presence in the moss bed of macrofossils of plants from different habitats, such as the occurrence of the moss *Scorpidium turgescens*, which frequently grows emergent from shallow water, with leaves of *Dryas integrifolia*, a plant of drier habitats.

Cushing (1964) notes that etched and pitted pollen grains, presumably corroded by the activity of aerobic bacteria, can be suspected of having been redeposited penecontemporaneously. Grains corroded in this manner have been found in our samples, but the incomplete knowledge of the processes and significance of pollen degradation makes their presence difficult to evaluate. The only safe conclusion seems to be that percentages of certain taxa in our pollen diagram may not be true indications of the vegetational importance of these taxa at the time the sediments accumulated.

We feel also that the fluctuations in the percentage of spruce, pine, and Cyperaceae pollen reflect redepositional phenomena rather than actual changes in the surrounding vegetation. It is tempting to attribute the decrease of Cyperaceae upward in the diagram to the rise in spruce and pine pollen, indicating perhaps an increased development of tree-dominated vegetation near the site. Ecologically this is reasonable because the same sequence appears repeatedly in late-glacial deposits in the Great Lakes region. Even though we have no direct evidence about the length of time required to accumulate the bryophyte bed and the lacustrine sediments directly above, it was probably not longer than a century, and perhaps no more than several decades. This is sufficient time for many vegetational changes to occur, but it seems unlikely that a threefold increase in spruce would have taken place during this period.

It is doubtful, however, that redeposition seriously changes our original interpretation of the vegetation. If the pollen in the till sample is a correct indication of what can be expected to be overrepresented in our spectra, then the percentage of tree pollen must be reduced. This eliminates most of the deciduous tree taxa and substantially decreases the amount of *Pinus* and *Picea*. A less sizable reduction in the percentages of various components of the nonarboreal category is necessary as well. This leaves us with spectra still dominated largely by herbs. The macrofossils are further evidence that the vegetation was open and herb- or shrub-dominated. No macrofossils of higher plants characteristic of the boreal forest were found in the deposit. Although this does not preclude elements of the boreal forest being located at some distance from the site, they certainly were not nearby.

The search for an analogue of the vegetation indicated by the fossils is hampered by the imperfect knowledge of arctic and subarctic plant communities and their pollen rain. The surface lake mud samples collected at sites surrounded by tundra vegetation near Barrow and Umiat on the arctic slope of Alaska that were published by Livingstone (1955) contain only traces of *Picea*, much higher Gramineae, and generally less pollen of other herbs than our spectra show. His surface samples from lakes farther south but still north of the Arctic Circle agree more closely except for the substantial amounts of *Alnus* and *Betula* in his samples and their absence from ours. It is well-known, of course, that *Alnus* and *Betula* do not have important expression in late-glacial pollen diagrams from the Great Lakes region.

The extensive series of surface samples from the Canadian arctic and subarctic recently published by Ritchie and Litchi-Federovich (1967) are somewhat more helpful.

The samples were obtained from five major vegetation types ranging from the high-arctic rock desert and fell-field southward through the sedge-moss tundra, the dwarf-shrub tundra, the forest-tundra ecotone, to the boreal forest. A perfect match between our spectra and theirs is not to be expected, but ours seem most closely related to the pollen rain at stations in the dwarf-shrub tundra or the forest-tundra ecotone. The amount of spruce in our samples, which may be overrepresented, suggests a closer relation to stations in the tundra-forest intergrade.

Spectra similar to ours have also been found at many sites in north-central and northeastern United States in sediments that accumulated immediately after retreat of the ice. The herb pollen assemblage zone of this time has generally been interpreted as representing either tundra or a mixture of tundra and boreal forest communities. Of the late-glacial pollen spectra reviewed by Davis (1967), ours resemble most closely those from southern New England. Both have large amounts of nonarboreal pollen, but most of our spectra have greater percentages of spruce. The large nonarboreal-pollen content of the New England spectra has been taken to indicate either that forest was far removed from the sites of pollen deposition or that the environment favored the production of large amounts of herb pollen (Davis, 1967).

An analysis of the habitats occupied by living counterparts of the macrofossils allows the reconstruction of several plant communities at and immediately around the site. Before the upper till covered the area, the sand beneath the bryophyte bed was at the surface. It must have provided a mosaic of well- and poorly drained sites available for plant growth. The abundant remains of the moss *Scorpidium turgescens* suggest that the sample studied for macrofossils was taken from one of the wetter areas, perhaps a shallow depression characterized by fluctuating water level. Such depressions were undoubtedly rich in calcium carbonate leached from the surrounding drift. Higher and better drained sites were probably less calcareous due to the removal of carbonates by runoff water or by percolation. One or several aquatic species of *Carex* may have grown with the *Scorpidium*, although we have no macroscopic evidence of their presence. *Campylium stellatum, Drepanocladus aduncus* var. *polycarpus*, and *D. revolvens* var. *intermedius* also belong to this community, although they probably occupied the drier margins of the depression. This is a type of rich fen vegetation (Sjörs, 1961, p. 14–18).

On the better drained slopes above, any water accumulated in the depression would have grown *Catoscopium nigritum* and *Ditrichum flexicaule. Bryum cryophilum* probably occupied a similar habitat in this area, but one kept moist by continuous seepage. *Bryum pseudotriquetrum* may have inhabited both the edge of the depression and drier sites above it. Associated with these mosses on the slopes above the depression were *Carex supina, Dryas integrifolia, Salix herbacea*, and *Vaccinium uliginosum* var. *alpinum* and perhaps other *Carex* species. *Carex tenuiflora* and/or *C. trisperma* perhaps grew at the edge of the depression. It is probable that islands of spruce trees were scattered across the landscape but that few were close to the bryophyte bed.

The physiognomy of the bryophyte- and herb-dominated vegetation probably was similar to that found on outwash plains in regions of present-day glaciation. This does not imply, however, that the ice front was necessarily close to the organic bed during its formation. No data are available that permit the determination of this distance.

In summary, we conclude that:

(1) The pollen diagram and macrofossils establish the occurrence of at least two plant assemblages that were present in northwestern Cheboygan County, Michigan, sometime between 12,500 and 13,300 years B.P.: a wet-site community dominated by mosses and a drier site community of other moss species and tundra species of vascular plants, including *Carex supina, Dryas integrifolia, Salix herbacea*, and *Vaccinium uliginosum* var. *alpinum*. A third community in which spruce trees occurred is also indicated by the

pollen diagram, but it was probably not represented in the immediate vicinity of the site and may have been no closer than some miles away.

(2) These communities are represented by fossils of species that today are characteristic of the northern boreal coniferous forest, the forest-tundra ecotone, and the tundra adjacent to the limit of trees, as in the vegetation belts now existing east and west of Hudson Bay.

(3) It can be inferred that the environment at the site during the time the bryophyte bed was accumulating had much in common with the environments of areas where the ranges of these plants coincide today—somewhere in the low arctic or the tundra-boreal forest ecotone—although the late-glacial ice-laden and shifting soils and the pioneer status of the vegetation must be kept in mind as modifying influences that leave certain doubts in this interpretation.

REFERENCES CITED

Argus, G. W., and Davis, Margaret B., 1962, Macrofossils from a late-glacial deposit at Cambridge, Massachusetts: Am. Midland Naturalist, v. 67, p. 106–117.

Baker, R. G., 1965, Late-glacial pollen and plant macrofossils from Spider Creek, southern St. Louis County, Minnesota: Geol. Soc. America Bull., v. 76, p. 601–609.

Bamberg, S. A., and Pemble, R. H., 1968, New records of disjunct arctic-alpine plants in Montana: Rhodora, v. 70, p. 103–112.

Barendsen, G. W., Deevey, E. S., and Gralenski, L. J., 1957, Yale natural radiocarbon measurements III: Science, v. 126, p. 908–919.

Benninghoff, W. S., 1960, Pollen spectra from bryophytic polsters, Inverness Mud Lake Bog, Cheboygan County, Michigan: Michigan Acad. Sci., Arts, and Letters Papers, v. 45, p. 41–60.

Black, R. F., Hole, F. D., Maher, L. J., and Freeman, Joan E., 1965, The Wisconsin Quaternary, p. 56–81 in Guidebook for field conference C, Upper Mississippi Valley, Assoc. Quaternary Research, VII Congress: Lincoln, Nebraska Acad. Sci.

Broecker, W. S., and Farrand, W. R., 1963, Radiocarbon age of the Two Creeks forest bed, Wisconsin: Geol. Soc. America Bull., v. 74, p. 795–802.

Cheney, L. S., 1931, More fossil mosses from Wisconsin: Bryologist, v. 34, p. 93–94.

Conard, H. S., 1938, The foray of 1938: Bryologist, v. 41, p. 139–142.

Crum, H., 1964, Mosses of the Douglas Lake region of Michigan: Michigan Botanist, v. 3, p. 3–12, 48–63.

Crum, H., Steere, W. C., and Anderson, L. E., 1965, A list of the mosses of North America: Bryologist, v. 68, p. 377–432.

Cushing, E. J., 1964, Redeposited pollen in late-Wisconsin pollen spectra from east-central Minnesota: Am. Jour. Sci., v. 262, p. 1075–1088.

——1967, Late-Wisconsin pollen stratigraphy and the glacial sequence in Minnesota, p. 59–88 in Cushing, E. J., and Wright, H. E., Jr., Editors, Quaternary paleoecology: New Haven, Connecticut, Yale Univ. Press, 433 p.

Darlington, H. T., 1964, The mosses of Michigan: Cranbrook Inst. Sci. Bull. 47, 212 p.

Davis, Margaret B., 1967, Late-glacial climate in northern United States: a comparison of New England and the Great Lakes region, p. 11–43 *in* Cushing, E. J., and Wright, H. E., Jr., *Editors*, Quaternary paleoecology: New Haven, Connecticut, Yale Univ. Press, 433 p.

Dickson, J. H., 1967, The British moss flora of the Weichselian glacial: Rev. Palaeobotany and Palynology, v. 2, p. 245–253.

Faegri, K., and Iversen, J., 1964, Textbook of pollen analysis: Copenhagen, Munksgaard, 237 p.

Fernald, M. L., 1950, Gray's manual of botany, 8th edition: New York, American Book Co., 1632 p.

Hultén, E., 1959, Studies in the genus *Dryas*: Svensk Bot. Tidskr., v. 53, p. 507–542.

―― 1962, The circumpolar plants. I. Vascular cryptogams, conifers, monocotyledons: Kungl. Svenska Vetenskapsakad. Handl., Fjärde Serien, v. 8, no. 5, 275 p.

Janssen, C. R., 1966, Recent pollen spectra from the deciduous and coniferousdeciduous forests of northeastern Minnesota: a study in pollen dispersal: Ecology, v. 47, p. 804–825.

Lawrence, D. B., Schoenike, R. E., Quispel, A., and Bond, G., 1967, The role of *Dryas drummondii* in vegetation development following ice recession at Glacier Bay, Alaska, with specific reference to its nitrogen fixation by root nodules: Jour. Ecology, v. 55, p. 793–813.

Livingstone, D. A., 1955, Some pollen profiles from arctic Alaska: Ecology, v. 36, p. 587–600.

Mårtensson, O., 1956, Bryophytes of the Torneträsk area, northern Swedish Lappland. II: Musci. Kungl. Svenska Vetenskapsakad. Avh. Naturskyddsarenden, no. 14, 321 p.

Persson, H., and Sjörs, H., 1960, Some bryophytes from the Hudson Bay lowland of Ontario: Svensk Bot. Tidskr., v. 54, p. 247–268.

Polunin, N., 1948, Botany of the Canadian eastern arctic. Part III. Vegetation and ecology: Natl. Mus. Canada Bull. 104, 304 p.

―― 1959, Circumpolar arctic flora: Oxford, The Clarendon Press, 514 p.

Porsild, A. E., 1957, Illustrated flora of the Canadian Arctic Archipelago: Natl. Mus. Canada Bull. 146, 209 p.

――1958, Geographical distribution of some elements in the flora of Canada: Geographical Bull. no. 11, p. 57–77.

Ritchie, J. C., and Litchi-Federovich, Sigrid, 1967, Pollen dispersal phenomena in arctic-subarctic Canada: Rev. Palaeobotany and Palynology, v. 3, p. 255–266.

Roberts, Patricia R., 1965, New records of arctic species in southeastern New Brunswick: Rhodora, v. 67, p. 92–93.

Rubin, M., and Alexander, Corrinne, 1958, U.S. Geological Survey radiocarbon dates IV: Science, v. 127, p. 1476–1487.

Schofield, W. B., and Robinson, H., 1960, Late-glacial and postglacial plant macro-fossils from Gillis Lake, Richmond County, Nova Scotia: Am. Jour. Sci., v. 258, p. 518–523.

Sjörs, H., 1961, Forest and peatland at Hawley Lake, northern Ontario: Natl. Mus. Canada Bull. 171, p. 1–31.

Spurr, S. H., and Zumberge, J. H., 1956, Late Pleistocene features of Cheboygan and Emmet Counties, Michigan: Am. Jour. Sci., v. 254, p. 96–109.

Steere, W. C., 1947, Musci, p. 370–490 in Polunin, N., Botany of the Canadian eastern Arctic, Part II: Thallophyta and Bryophyta: Natl. Mus. Canada Bull. 97.

Suess, H. E., 1954, U.S. Geological Survey radiocarbon dates I: Science, v. 120, p. 467–473.

Terasmae, J., 1960, A palynological study of the Pleistocene interglacial beds at Toronto, Ontario, p. 23–40 in Terasmae, J., Contributions to Canadian paly-nology no. 2: Geol. Survey Canada Bull. 56.

Watts, W. A., 1967, Late-glacial plant macrofossils from Minnesota, p. 89–97 in Cushing, E. J., and Wright, H. E., Jr., Editors, Quaternary paleoecology: New Haven, Connecticut, Yale Univ. Press, 433 p.

Watts, W. A., and Winter, T. C., 1966, Plant macrofossils from Kirchner Marsh, Minnesota—a paleoecological study: Geol. Soc. America Bull., v. 77, p. 1339–1359.

West, R. G., 1961, Late- and postglacial vegetational history in Wisconsin, particu-larly changes associated with the Valders readvance: Am. Jour. Sci., v. 259, p. 766–783.

GEOLOGICAL SOCIETY OF AMERICA, INC.

SPECIAL PAPER 123

Cary–Port Huron Interstade: Evidence from a Buried Bryophyte Bed, Cheboygan County, Michigan

WILLIAM R. FARRAND, ROBERT ZAHNER, AND WILLIAM S. BENNINGHOFF

University of Michigan, Ann Arbor, Michigan

ABSTRACT

Radiocarbon dating of a thin layer of mosses sandwiched between two thick layers of reddish-brown sandy clay till yielded an age of 12,500 to 13,000 years B.P. These dates and a pollen spectrum that suggests a locally treeless flora form the basis for correlation with the Cary–Port Huron interstade of the Wisconsin glaciation. The absence of a Two Creeks horizon at this site and the presence of a single unit of till showing no evidence of an erosional break between Cary–Port Huron time and Glacial Lake Algonquin time is interpreted as evidence that the Straits of Mackinac was not deglaciated during Two Creeks time. It would have been the Petoskey-Cheboygan lowland, according to this interpretation, that drained the low-level Two Creeks lake from Lake Michigan to Lake Huron. Furthermore, in northern Cheboygan County red clayey till, formerly considered to be unique in this area to the Valders Stade, was deposited also by Port Huron and pre–Port Huron ice.

CONTENTS

INTRODUCTION

The most recent advance of continental ice sheets into the Southern Peninsula of Michigan is attributed to the Valders Stade of the Wisconsin glaciation (Fig. 1). The characteristic red, clayey till of the northern part of the Southern Peninsula has been correlated with the type Valders drift of northeastern Wisconsin by several workers (Melhorn, 1954; Spurr and Zumberge, 1956). The Valders advance in the state of Wisconsin was marked by the overriding of the now-famous Two Creeks forest (Thwaites and Bertrand, 1957), wood of which has been dated by the radiocarbon method as 11,840 ± 100 years old (Broecker and Farrand, 1963). The overriding of the Two Creeks forest occurred only a short time, perhaps a few hundred years, before the Valders ice reached its maximum extent near Milwaukee, 85 miles south of the Two Creeks locality.

There is no direct date for the Valders Stade east of Lake Michigan. In the investigation reported here we believed at first that we were dealing with a stratigraphic situation parallel to that at Two Creeks. The radiocarbon dates, however, suggested that this was not the case, and further complications came to light when we investigated the deposits underlying the bryophyte bed described below. In consideration of these problems we have concluded that (1) the Cheboygan County deposits represent an oscillation of the ice front prior to the Two Creeks oscillation, and (2) it is possible that the Straits of Mackinac was not deglaciated during the Two Creeks Interstade.

ACKNOWLEDGMENTS

Much credit is due E. Ginop, on whose farm the bryophyte bed is located, for bringing this discovery to our attention, for permission to excavate for uncontaminated samples, and for his assistance in the excavation and core drilling. Core drilling in 1967 was financed by the Institute of Science and Technology of the University of Michigan. The mosses were identified by Dr.

H. A. Crum, curator of bryophytes at the University of Michigan Herbarium. The radiocarbon dates reported herein were run through the courtesy of D. L. Thurber (Lamont Geological Observatory), James B. Griffin (University of Michigan's Michigan Memorial–Phoenix Project), and Meyer Rubin (U.S. Geological Survey).

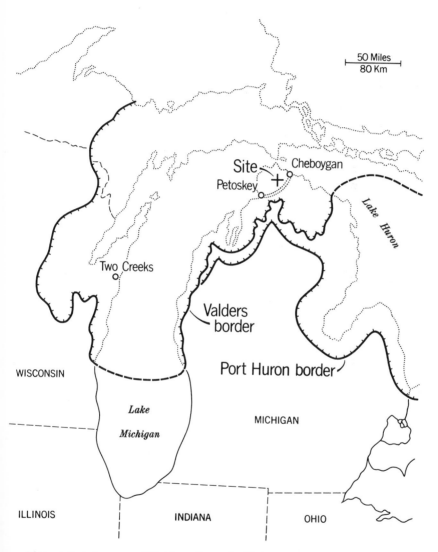

Figure 1. Location map of Michigan. Shows the Cheboygan County bryophyte locality, Two Creeks, Wisconsin, the maximum extent of the continental ice sheet during the Valders and Port Huron Stades, and the Petoskey-Cheboygan lowland (double dotted line).

SITE DESCRIPTION

Geographic Setting

In August 1965, during the excavation of a farm pond in Cheboygan County, Michigan, a fossil bryophyte community was uncovered and brought to our attention. This site is on the Ginop property near the exact center of Section 4, Munro Township (T.37N., R.3W.), approximate latitude 45° 38′ N., longitude 84° 41′ W., at an altitude of 226 m (740 feet) above sea level, or 49 m (160 feet) above the level of Lake Michigan. The surficial geology of Cheboygan County has previously been mapped in detail by Spurr and Zumberge (1956), who concluded that the surface drift at the site is of Valders age. At the site about 3 m of red, clayey Valders-type till was found above the bryophyte bed.

Stratigraphy

The stratigraphy of the site is shown in Figures 2 and 3. Surface probing showed the areal extent of the bryophyte bed to be roughly 400 m². An excavation 2 m × 4 m × 4.5 m deep was made in 1965 next to the pond (Fig. 3) in order to examine the stratigraphy and to obtain uncontaminated samples for identification, radiocarbon dating, and pollen analysis. Ground water was encountered at a depth of 2 m, which restricted the extent of the excavation and necessitated pumping as the digging progressed. Core drilling was undertaken in November 1967 in order to investigate the stratigraphy below the bryophyte bed. Four holes were drilled, and three of them encountered organic material associated with sandy sediments at the same depth as indicated by the earlier excavation (Figs. 2, 3). The absence of both organic material and sand in core hole #1 confirms the opinion of Mr. Ginop and his associates that the peaty material underlies only the southern part of the pond area.

Core drilling, however, introduced another complication. If the bryophyte bed is correlative with the Two Creeks forest bed, as was assumed at first, then the underlying deposits should be pre-Valders in age. In this area of Michigan pre-Valders deposits are represented by Port Huron till. Melhorn (1954, p. 142–151) was able to distinguish the blue or gray Port Huron ("Cary") till from the younger, red Valders till throughout much of the northern part of the Southern Peninsula of Michigan, especially in the area somewhat to the south of Cheboygan County. Thus, we expected to find blue or gray till below the bryophyte bed and were quite surprised to find more red clayey till, identical to that overlying the bryophyte bed in both color (Fig. 3) and texture (Fig. 4). There are at least 7 m (23 feet) of red till below the organic layer. (The drilling operation was limited to 35-foot penetration.) This discovery suggests that in the northernmost part of southern Michigan either the typical red-till-over-gray-till stratigraphy does not apply or that the bryophyte bed does not represent the Port Huron–Valders (Two Creeks) interstade, or both. We will return to this question later.

Comparison of Red Tills

A lithological comparison of these tills with local surface ("Valders") tills analyzed by Melhorn (1954, p. 137, 149) shows that the colors are identical (Munsell notation 5 YR 4–5/3, moist, reddish brown) but that the grain size is somewhat coarser (Fig. 4) than Melhorn's values for seven or eight "typical" Valders samples. The principal difference is the inclusion of 15 to 20 percent

Cheboygan Co. Bryophyte Bed

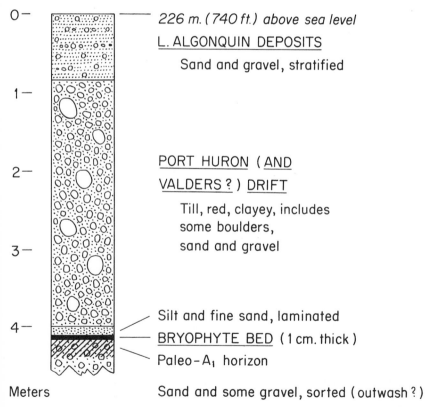

0 —

226 m. (740 ft.) above sea level

L. ALGONQUIN DEPOSITS

Sand and gravel, stratified

1 —

2 —

PORT HURON (AND

VALDERS ?) DRIFT

Till, red, clayey, includes some boulders, sand and gravel

3 —

4 —

Silt and fine sand, laminated

BRYOPHYTE BED (1 cm. thick)

Paleo-A$_1$ horizon

Meters

Sand and some gravel, sorted (outwash ?)

Figure 2. Stratigraphy of the Cheboygan County bryophyte locality based on 1965 excavation.

more medium and coarse sand in our samples than in those of Melhorn. Only our sample 2–35′ agrees well with Melhorn's data.

We also compared the tills from the bryophyte locality with a long series of granulometric analyses of Valders drift from Wisconsin (Lee and others, 1962, p. 155). Here again the Cheboygan County samples (with the exception of sample 2–35′) are distinctly coarser than the great majority of Valders tills. Four of our samples are sandy loams, in contrast to the usual Valders till of Wisconsin, which is classed as clay, silty clay, or clay loam. A tentative conclusion that can be drawn from this limited comparison is that red Valders till is generally richer in clay than the red pre-Valders tills of the same area.

The Bryophyte Bed

The bryophyte bed is essentially a mat of mosses, highly compressed to a thickness of about 1 cm. This moss community was growing on the surface of

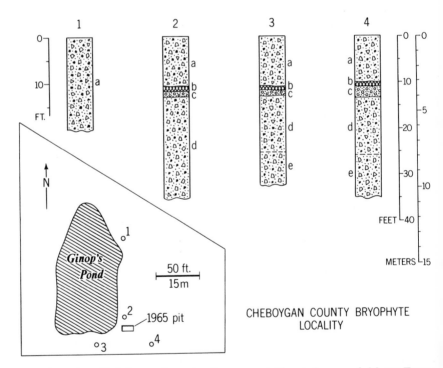

Figure 3. Driller's logs and core locations around Ginop's farm pond, Munro Township, Cheboygan County, Michigan. 1967 core locations are shown by numbered circles. The original 1965 stratigraphic excavation is shown by the rectangular outline. The stratigraphic units are: (a) upper red sandy clay till, (b) gray sand with organic material, (c) brown sand with some gravel, (d) lower red sandy clay till, (e) same as (d) with sandy streaks and more cobbles.

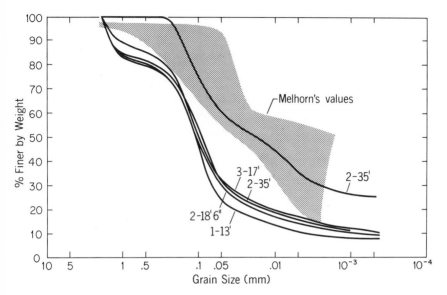

Figure 4. Cumulative granulometric curves of red tills above and below the Cheboygan bryophyte bed compared with those of Valders red till by Melhorn (1954, p. 149). Sample numbers refer to core number and depth in core (*compare* Fig. 3).

a sand and gravel deposit which resembles glacial outwash. A grayish paleo-A_1 horizon of a fossil soil had developed in the upper part of the sand and gravel immediately below the bryophyte bed. The mosses are covered by laminated fine sand and silt, some of which appears to be varved; this unit is no more than 10 cm thick. This important complex is deeply buried beneath about 3 m of red, clayey till (described above). At the surface of the section there is about 0.8 m of stratified sand and gravel that is continuous laterally with deposits of the beach of Glacial Lake Algonquin. The Algonquin shoreline lies about 150 m to the northeast of the site and is only a few meters higher in altitude than the site, thus about 227 to 228 m (740 to 745 feet) above sea level.

The megascopic plant material is nearly 100 percent mosses, including eight species; *Scorpidium turgescens* (Th. Jens.) Loeske and *Catoscopium nigritum* (Hedw.) Brid. are the most abundant. These mosses today commonly occupy low, wet fen communities in the boreal forest region. Both species occur at present in northern Michigan, but the center of their range is boreal Canada. Among the recognizable fragments of plant debris separated and identified by Norton Miller of Michigan State University were leaves of the arctic and alpine tundra species *Dryas integrifolia* (N. Miller and W. S. Benninghoff, this volume).

Pollen Analysis

Seven samples available for pollen analysis all proved to hold very low concentrations of pollen and spores. The richest samples contain almost exclusively spruce (*Picea*) and sedges (Cyperaceae). There are possibly two species of spruce present but apparently only one species of Cyperaceae; this resembles the pollen of *Eriophorum* (cottongrass), but firm identification of the genus is impossible. There is a very small proportion of balsam fir (*Abies*) pollen and approximately equal amounts of pine (*Pinus*) pollen. There are also trace proportions of larch (*Larix*), oak (*Quercus*), and grass (Gramineae) pollen. A large proportion of the pine, larch, and oak pollen is damaged and is possibly redeposited material.

The low concentration of pollen and spores could be attributed either to rapid deposition that diluted the pollen contributed to the sediment or to low concentrations of pollen in the atmosphere over the site because of sparse vegetation. In one of the samples the concentration of Cyperaceae pollen is the highest ever seen by Benninghoff, a palynologist, exceeding that of sedge marsh sediments. This would imply that the sedimentation rate was not greater than usual in sedge marshes. Yet the concentration of the total tree-pollen load in the atmosphere was very greatly less than in circumstances where these trees are common, even in very open stands, in the landscape around the site of deposition. We believe these sediments may record a preforest stage in plant succession on this site and that the nearest forest, one of spruce with some balsam fir, may have been on the order of 80 km (50 miles) away.

It is interesting to compare this pollen spectrum with that of the Two Creeks site in Wisconsin. A recent palynological study of the type Two Creeks forest bed (West, 1961, p. 771) shows that spruce (both *Picea mariana* and *P. glauca*) account for 70 to 90 percent of the total pollen present except for the lowest sample in the 14-cm-thick section, which contains only 5 percent spruce and 95 percent *Shepherdia canadensis* (the soapberry bush). Cyperaceae never rise above 5 percent of the pollen sum at any level in West's pollen diagram.

On the other hand, Schweger (*in* International Association for Quaternary Research, 1965, p. 66–68) has prepared a pollen diagram from the Duck Creek Ridge site, 12 km southwest of Green Bay, that is correlated with the Two Creeks section. In Schweger's diagram sedge pollen decreases from 80 percent of the total pollen at the bottom of the diagram to about 20 percent at the top, and spruce increases from about 1 percent to nearly 80 percent from bottom to top.

It thus appears that the flora of the Cheboygan County site was distinctly different from that of the type Two Creeks site but similar to the earlier portion of the Duck Creek sequence. The treeless vegetation in Cheboygan County is believed to represent an earlier stage in the plant succession due to its being older in time than the Two Creeks flora or to its closer proximity to the front of the ice sheet, or both.

Radiocarbon Dating

The bryophyte bed was dated by the Lamont, Michigan, and U.S. Geological Survey radiocarbon laboratories. The Lamont laboratory subjected the sample to treatment by hot alkali and acid to remove any possible contaminants and also made an extraction of humic material, which was dated separately. The Washington and Lamont dates[1] are given in Table 1.

TABLE 1. RADIOCARBON DATES FROM CHEBOYGAN COUNTY BRYOPHYTES

Sample no.	Date (years B.P.)
W-1847	12,500 ± 500
W-1889	12,570 ± 500
L-1064 (alkali-acid)	13,300 ± 400
L-1064 (humic)	12,800 ± 400

The age of the Cheboygan bryophyte bed thus appears to be between 12,500 and 13,000 years, that is, about 1000 years older than the Two Creeks forest.

STRATIGRAPHIC INTERPRETATION

The most obvious interpretation of the site stratigraphy is that the bryophyte bed is a remnant of an *in situ* moss community covered by lacustrine silts of a lake ponded by the advancing ice sheet and subsequently overridden by that ice sheet and covered with the red till. Following the retreat of the ice sheet from this locality, the site was inundated by the waters of Glacial Lake Algonquin.

The possibility that the bryophyte bed was covered with red till by solifluction or other mass movement was also considered. This interpretation was rejected, however, mainly because the site is separated from higher land by some 150 m of flat terrain.

As stated above, the stratigraphy of this site is very nearly identical to that of the type Two Creeks site in Wisconsin. At Two Creeks spruce trees were growing during a relatively low stage of Lake Michigan (less than 175 m [580 feet] above sea level). About 11,800 years ago these trees were killed by rising lake waters and partially covered by lacustrine sand. This water-level rise has been correlated with the advance of the Valders ice which blocked the Straits of Mackinac. A short time later the Valders ice overrode the dead forest at Two Creeks, covering it with red, clayey till.

---

[1] Another date of 9960 ± 350 years B.P. (sample M-1753, Crane and Griffin, 1968, p. 64) obtained from this same material does not accord with the Washington and Lamont dates. The Michigan date suggests that the bryophytes are *younger* than both the Valders till and Glacial Lake Algonquin, and this contradicts the stratigraphic evidence. Although the reason for this discrepancy is not known, this date is rejected because the other dates show good agreement.

At the Cheboygan County site the organic layer is nearly 100 percent bryophytes; there are woody shrubs, but no trees, in contrast to Two Creeks. The bryophyte bed is covered by lacustrine silts and then by red, clayey till. Both at Two Creeks and in Cheboygan County the beach of Glacial Lake Algonquin is demonstrably younger than the red, clayey till.

The radiocarbon dates, however, show a discrepancy between the two sites. The Cheboygan County dates (12,500 to 13,300 years) are roughly 1000 years older than the most recently and most precisely determined Two Creeks dates (11,840 ± 100 years B.P.), which is the average of six different samples (Broecker and Farrand, 1963). Previously reported Two Creeks dates ranged between 10,400 and 12,200 years, although the standard deviation was quite large in some cases. Thus the Cheboygan County dates seem to lie distinctly outside the range of variation of the Two Creeks forest-bed dates. On the other hand, the Cheboygan County dates coincide exactly with previously reported dates (listed in Table 2) for the ice-front oscillation just prior to the Two Creeks oscillation, that is, the Cary–Port Huron (Lake Arkona) interstade. An ice-front retreat and readvance at this time has long been inferred from glacial-lake history (Hough, 1958, p. 146–149). The low Lake Arkona stage between preceding (Maumee) and following (Whittlesey) higher lake stages has been the primary evidence for withdrawal of the ice front along the axis of the Thumb of Michigan in mid-Cary time and a subsequent readvance which built the Port Huron moraine. The amount of horizontal displacement of the ice front in the Thumb area at the time of the Port Huron advance is not known. Furthermore, recent subsurface investigations of Lake Erie have suggested that the area of Niagara Falls was deglaciated during the Cary–Port Huron interval (Wall, 1968, p. 103). Still other evidence comes from the Huron River Valley in southeastern Michigan (Kunkle, 1963) and from the glacial stratigraphy in southern Ontario (Dreimanis, 1968).

In addition to the radiocarbon dates, the plant assemblage suggests that the Cheboygan County bryophyte bed is older than the Two Creeks

TABLE 2. RADIOCARBON DATES PERTINENT TO CARY–PORT HURON INTERSTADE

Sample no.	Dated event and reference	Date (years B.P.)
Y-240	Whittlesey beach at Bellevue, Ohio (Barendsen and others, 1957, p. 912)	12,800 ± 250
W-430	Peat below Whittlesey gravel, Parkerstown, Ohio (Rubin and Alexander, 1958, p. 1477)	12,920 ± 400
W-33	Transition from Arkona to Whittlesey, Cleveland, Ohio (Suess, 1954, p. 469)	13,600 ± 500

forest bed. The inferred treeless flora of Cheboygan County certainly represents an earlier stage in plant succession than the spruce forest at Two Creeks. However, the distance from each locality to the contemporaneous ice front, which is largely unknown, could account for the floral differences. In any case, the plant remains do not contradict the conclusion that the Cheboygan bryophyte bed dates from the Cary–Port Huron interstade.

Accepting the above conclusion leads to another problem. If the bryophyte bed is Cary–Port Huron in age, then the overlying red clayey till is most likely of Port Huron age since there is no evidence whatsoever of erosion or of soil formation in the stratigraphic interval between the bryophyte bed and the red till. These two units are separated by only 10 cm of lacustrine deposits, and it is unlikely that they represent an interval of 1000 years or more, which would be the case if the red till just above the bryophyte bed were Valders in age.

Furthermore, there is but a single unit of till separating the bryophyte bed and Glacial Lake Algonquin deposits, which are in part contemporary with and in part postdate the Valders Stade (Hough, 1966, Tables 4, 5; Farrand, 1962, p. 185; Crane and Griffin, 1966, p. 258, sample M-1603). Where then are the deposits of Valders and Two Creeks ages? According to the usual reconstruction, the ice front retreated to a position north of the Straits of Mackinac during the Two Creeks Interstade, a position which allowed the level of ancestral Lake Michigan to drop to at least its present level and probably lower. If this were so, then erosion, nonglacial deposition, or soil formation should have occurred at that time everywhere south of the Straits, including Cheboygan County. The absence of Two Creeks deposits and of an erosional unconformity requires an explanation, and several possibilities come to mind.

First, it is possible that the entire 3-m-thick till section above the bryophyte bed is of Port Huron age. The drumlinized topography around the site suggests that there was a period of erosion by the ice sheet that followed the deposition of the red till. Accordingly, it can be suggested that that period of glacier erosion was related to the Valders ice activity and that there was no till deposition locally during the Valders Stade.

Another explanation suggests that the 3 m of red till above the bryophyte bed represents continuous deposition throughout Port Huron and Valders time, without a break during the Two Creeks Interstade. This hypothesis raises the question of the deglaciation of the Straits of Mackinac during Two Creeks time. There is another possible route, however, by which the Lake Michigan basin waters could have been drained to the Lake Huron basin. This route runs through the lowland from the head of Little Traverse Bay at Petoskey to the Lake Huron shore at Cheboygan (Fig. 1). This lowland is now occupied by a chain of lakes whose water surfaces are now no more than about 7 m (20 to 22 feet) above the level of Lakes Michigan and Huron. The only barrier in the lowland at the present time is the dune field at Petoskey that formed in relation to the much younger Nipissing shoreline (4000 to 5000 years B.P.). At Petoskey the Glacial Lake Algonquin shoreline has experienced about 30 m (100 feet) of rebound. Thus during the Two Creeks Interstade this lowland was 30 m or more lower than it is now, and it would easily have accommodated drainage from Petoskey to Cheboygan. With the functioning of this low-level route in Two Creeks time, there is no longer the necessity for the deglaciation of the Straits of Mackinac at that time. In fact, this second interpretation of the bryophyte-bed stratigraphy implies that the Straits was not opened at Two Creeks time. The continuous section of till from Cary–Port Huron time until Lake Algonquin time suggests that northern Cheboygan County was continuously covered by the ice sheet throughout this time period.

The succession of events according to this second interpretation would be as follows:

(1) Cary ice, after depositing red, sandy clay till, retreats northward across Cheboygan County;

(2) the bryophyte bed grows on outwash sand;

(3) the advancing Port Huron ice ponds water locally, flooding the mosses;

(4) Port Huron ice advances to its maximum position about 50 km (30 miles) south of the site and spreads red, sandy clay till over the bryophyte bed;

(5) during the Two Creeks Interstade the ice front withdraws northward to southern Emmet and central Cheboygan Counties, just north of the Petoskey-Cheboygan lowland, and Lake Michigan water drains to the Lake Huron basin;

(6) the Two Creeks forest grows in Wisconsin; an equivalent event is not yet known in Michigan;

(7) Valders ice advances to cover the Two Creeks forest in Wisconsin, perhaps 250 km (150 miles) advance or more along the axis of Lake Michigan, but only a minor readvance of some 40 km (25 miles) in the northern part of the Southern Peninsula of Michigan, the ice sheet being limited in its southern extent by the high crest of the Port Huron moraine;

(8) Valders ice withdraws, deglaciating the bryophyte locality, and, concomitantly, Glacial Lake Algonquin expands, covering a great percentage of the surface of the Southern Peninsula of Michigan from the Petoskey-Cheboygan lowland north.

In summary, three possible interpretations have been considered, namely, that the continuous section of red till between the bryophyte bed and Glacial Lake Algonquin deposits includes (1) Valders till only, (2) Port Huron till only, or (3) both Port Huron and Valders till, deposited without interruption.

The absence of evidence of erosion or weathering and of a significant thickness of deposits between the bryophyte bed and the base of the red till seems to rule out possibility 1. Possibility 2 would require nondeposition by Valders ice in this area, although Valders till has been recognized well south of the site area. Drumlinization in central Cheboygan County could represent the only recorded activity of the Valders ice sheet here. This hypothesis cannot be tested at present, and it does not give us any information on the amount of ice retreat between the Port Huron and Valders Stades. We favor possibility 3 as the simplest interpretation. It accords with all the known field observations. It implies that the Straits of Mackinac was not opened during Two Creeks time, but the Straits of Mackinac is no longer required as an outlet of a low-level lake in the Lake Michigan basin once the function of the Cheboygan-Petoskey lowland is recognized.

CONCLUSIONS

In summary, it appears that the Cheboygan County bryophyte bed is a remnant of the Cary–Port Huron (Lake Arkona) interstade. This conclusion is based on radiocarbon dating (12,500 to 13,000 years B. P.) and is supported by floral analysis (locally treeless flora). If this interpretation is correct, this bryophyte bed provides firm stratigraphic evidence for an interstade that is becoming increasingly well-documented. In addition, it implies that the Port Huron ice advanced at least 50 km (30 miles) in this part of Michigan.

A second conclusion is that both Port Huron till and pre–Port Huron till in north-central Cheboygan County are red clayey tills lithologically similar to much typical Valders till. This is the first confirmation of the occurrence of pre-Valders red tills in southern Michigan, although similar situations have been known for some time in Wisconsin (Frye and others, 1965, p. 57; Black, 1966, p. 170) and Minnesota (Wright and Ruhe, 1965, p. 36–38).

Third, the occurrence of a single body of till between the bryophyte bed and Glacial Lake Algonquin deposits (*circa* 11,500 years B.P.) is interpreted as evidence that the ice sheet did not retreat from northern Cheboygan County during the Two Creeks Interstade. This implies that the Straits of Mackinac was not deglaciated at that time and that Two Creeks drainage took place through the Petoskey-Cheboygan lowland.

REFERENCES CITED

Barendsen, G. W., Deevey, E. S., and Gralenski, L. J., 1957, Yale natural radiocarbon measurements III: Science, v. 126, p. 908–919.

Black, R. F., 1966, Valders glaciation in Wisconsin and upper Michigan—a progress report: Michigan Univ., Great Lakes Research Div. Pub. 15, p. 169–175.

Broecker, W. S., and Farrand, W. R., 1963, Radiocarbon age of the Two Creeks forest bed, Wisconsin: Geol. Soc. America Bull., v. 74, p. 795–802.

Crane, H. R., and Griffin, J. B., 1966, University of Michigan radiocarbon dates XI: Radiocarbon, v. 8, p. 256–285.

—— 1968, University of Michigan radiocarbon dates XII: Radiocarbon, v. 10, p. 61–114.

Dreimanis, A., 1968, Cary-Port Huron Interstade in Eastern North America and its correlatives (Abstract): Geol. Soc. America Spec. Paper 115, p. 259.

Farrand, W. R., 1962, Postglacial uplift in North America: Am. Jour. Sci., v. 260, p. 181–199.

Frye, J. C., Willman, H. B., and Black, R. F., 1965, p. 43–61 in Wright, H. E., Jr., and Frey, D. G., Editors, The Quaternary of the United States: Princeton, New Jersey, Princeton Univ. Press, 922 p.

Hough, J. L., 1958, Geology of the Great Lakes: Urbana, Illinois, Illinois Univ. Press, 313 p.

—— 1966, Correlation of glacial lake stages in the Huron-Erie and Michigan basins: Jour. Geology, v. 74, p. 62–77.

International Association for Quaternary Research, 1965, Guidebook for Field Conference C, 126 p.

Kunkle, G. R., 1963, Lake Ypsilanti: a probable late Pleistocene low-lake stage in the Erie basin: Jour. Geology, v. 71, p. 72–75.

Lee, G. B., Janke, W. E., and Beaver, A. J., 1962, Particle size analysis of Valders drift in eastern Wisconsin: Science, v. 138, p. 154–155.

Melhorn, W. N., 1954, Valders glaciation of the Southern Peninsula of Michigan: Ann Arbor, Michigan, Michigan Univ., Ph.D. dissert., 178 p.

Rubin, M., and Alexander, C., 1958, U.S. Geological Survey radiocarbon dates IV: Science, v. 127, p. 1476–1487.

Spurr, S. H., and Zumberge, J. H., 1956, Late Pleistocene features of Cheboygan and Emmet Counties, Michigan: Am. Jour. Sci., v. 254, p. 96–109.

Suess, H. E., 1954, U.S. Geological Survey radiocarbon dates I: Science, v. 120, p. 467–473.

Thwaites, F. T., and Bertrand, K., 1957, Pleistocene geology of the Door Peninsula, Wisconsin: Geol. Soc. America Bull., v. 68, p. 831–879.

Wall, R. E., 1968, A sub-bottom reflection survey in the central basin of Lake Erie: Geol. Soc. America Bull., v. 79, p. 91–106.

West, R. G., 1961, Late- and postglacial vegetational history in Wisconsin, particularly changes associated with the Valders readvance: Am. Jour. Sci., v. 259, p. 766–783.

Wright, H. E., Jr., and Ruhe, R. V., 1965, p. 29–41 in Wright, H. E., Jr., and Frey, D. G., Editors, The Quaternary of the United States: Princeton, New Jersey, Princeton Univ. Press, 922 p.

GEOLOGICAL SOCIETY OF AMERICA, INC.
SPECIAL PAPER 123

A Stratigraphic Record of the Pollen Influx to a Lake in the Big Woods of Minnesota

JEAN C. B. WADDINGTON

Limnological Research Center,
University of Minnesota, Minneapolis, Minnesota

ABSTRACT

Determination of pollen concentration (absolute pollen frequency), combined with eight radiocarbon dates from a 16-m core, provides an estimate of pollen influx to a small lake in the present Big Woods area of south-central Minnesota. The observed variation in pollen influx is difficult to relate to the vegetational history inferred from the relative pollen stratigraphy. An alternative hypothesis relates the changes in pollen influx to variation in the water level and movement of sediment in the lake basin.

The pollen stratigraphy is consistent with other sites in Minnesota and South Dakota; it indicates that the Rutz Lake area was covered by boreal forest at the time the lake was formed and by a parkland of deciduous trees shortly after. Prairie dominated the area from about 8000 to 3200 years ago, when an increase in moisture favored an oak parkland again. The remaining openings were invaded by the mixed deciduous forest known as the Big Woods about 400 years ago.

An index of the charcoal content of the sediment is maximum during the time of greatest prairie expansion and confirms the importance of fire in the prairie environment.

CONTENTS

INTRODUCTION

The term "Big Woods" has been widely used to describe a large area of mixed deciduous forest occupying central Minnesota roughly between the Mississippi and Minnesota Rivers and between the mixed coniferous-deciduous forest to the northeast and the prairie to the southwest (Fig. 1). This investigation was planned to contribute to the understanding of the vegetational history of this area in relation to other parts of the state and to test some of the ideas of Daubenmire (1936) on the origin of the Big Woods. A lake in the heart of the area was chosen, and a stratigraphic study was made of its sediments to determine the variation with time in the relative and absolute frequency of preserved pollen and the concentration of charcoal fragments in the sediment.

ACKNOWLEDGMENTS

I am indebted to Dr. E. J. Cushing, who suggested this research topic and who has provided continued guidance and advice, and to Dr. H. E.

Wright, Jr., for the use of his laboratory facilities and for sustained encouragement and advice. Both are thanked for help in revising the manuscript. Dr. J. H. McAndrews contributed in useful discussion and allowed me to use some unpublished data. The radiocarbon dates were provided by M. Stuiver. E. J. Cushing, N. Potter, H. E. Wright, Jr., and C. J. Waddington are thanked for their assistance in obtaining the core.

The work was supported in part by funds granted to the University of Minnesota by the Hill Foundation (St. Paul) and by National Science Foundation Grants GB-3401 and GB-7163.

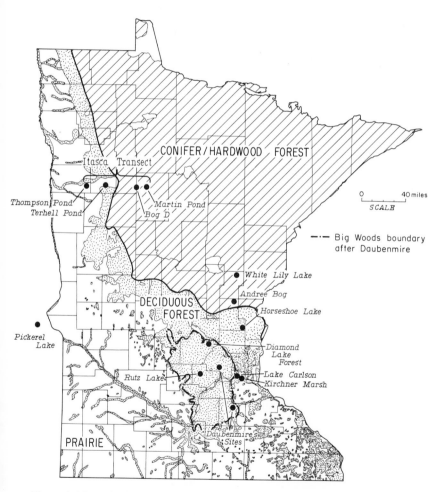

Figure 1. Map of Minnesota showing vegetation regions and sites mentioned in text. Vegetation *from* Upham (Winchell, 1884a). Big Woods boundary *from* Daubenmire (1936).

COMPOSITION AND KNOWN HISTORY OF THE BIG WOODS

When Nicollet (1841) made a reconnaissance of the area for the U.S. Government, he mapped and described the vegetation regions. According to him, "The French give to the forests the name of *Bois-francs*, or *Bois-forts*, whenever they are not composed principally of trees belonging to the family of the coniferae." Big Woods became the accepted translation of the French name. Winchell, in the county-by-county survey of Minnesota (1844b, 1888), refers to the deciduous forest of central Minnesota as the Big Woods and describes its distribution and composition. Table 1 lists the trees that he recorded from a county in the eastern part of the Big Woods. According to Winchell, the same tree species were present in most of the counties, but the dominant species differed. In some counties oak and elm dominated, in others elm and maple, or elm and basswood, or maple and basswood. The Big Woods appears to have been a complex mosaic of deciduous plant communities.

After about 1850, the forests of the Big Woods were rapidly cleared for agriculture. An attempt to reconstruct the boundaries and composition of the Big Woods from the scattered remnants was made by Daubenmire (1936). He described in detail two "near-virgin" stands in the eastern part of the area (Fig. 1). These were dominated by *Acer saccharum* (sugar maple) and *Tilia americana* (basswood), and Daubenmire concluded that the entire Big Woods was a continuous forest community dominated by these two species. According to a reconnaissance survey of the area, he produced a map of the boundaries of the community as it was at the time of settlement by white man. These boundaries are shown in Figure 1.

TABLE 1. TREE SPECIES IN HENNEPIN COUNTY IN ESTIMATED ORDER OF FREQUENCY*

Ulmus americana	*Carya cordiformis*
Tilia americana	*Amelanchier* sp.
Acer saccharum	*Pyrus* sp.
Quercus ellipsoidalis	*Larix laricina*
Juglans cinerea	*Acer negundo*
Acer saccharinum	*Populus deltoides*
Quercus macrocarpa	*P. grandidentata*
Acer rubrum	*Prunus serotina*
Populus tremuloides	*Carpinus caroliniana*
Fraxinus americana	*Salix nigra*
Quercus rubra	*Celtis occidentalis*
Ulmus rubra	*Betula papyrifera*
Ostrya virginiana	*Quercus alba*
Fraxinus nigra	*Juniperus virginiana*
Prunus americana	*Pinus strobus*

*After Winchell (1888, p. 277). Plant nomenclature altered to conform to Fernald (1950).

An additional maple-basswood stand in the region (the Diamond Lake stand, Fig. 1) was described by Bray (1956). The stands described by Daubenmire and Bray are very similar in structure and composition to the mesic southern hardwoods in Wisconsin described by Curtis (1959). Whether the entire Big Woods was ever as homogeneous in its upland vegetation as these stands, as Daubenmire implied, must remain an open question. Examination of the witness-tree data from notes of the Federal Land Survey in parts of the Big Woods by Janssen (1966, 1967) and McAndrews (1969) suggests that sugar maple and basswood were not the most important trees at the time of the survey. This observation has led McAndrews (1969) to propose that there has been succession to a more mesic forest during the 100 years since the survey.

RUTZ LAKE AND ITS SETTING

In the choice of a site for pollen analysis, the following criteria were used: (1) the lake should be near the center of the Big Woods area; (2) it should be small, to increase the chance that pollen of sugar maple and basswood would be represented in the sediment, for these trees disperse little pollen (Janssen, 1966); (3) it should be steep-sided or surrounded by upland on all sides, for the same reason; (4) it should be permanent within the knowledge of local residents; and (5) the sediment in the lake basin should be as uniform as possible, with no indication of interruptions in deposition. Of a number of lakes investigated, all but one were rejected because a layer of compact, clastic sediment occurred a few meters below the sediment surface. The layer suggested that the lake basins were severely eroded and that the lake may have dried up at some time in the past; pollen analysis of the sediments immediately above the clastic layer indicated its age to be mid-postglacial. The remaining site, Rutz Lake, did not have such a layer; it provided 16 m of remarkably homogeneous organic sediment.

Rutz Lake is in the SW.1/4NE.1/4, sec. 8, T.116N., R.25W., Carver County, Minnesota (Fig. 1). The lake is roughly oval, with an area of about 13 hectares; it occupies a shallow depression in gently rolling terrain that consists of gray calcareous till deposited by the Des Moines glacier lobe during a late phase of the Wisconsin glaciation (Wright and Ruhe, 1965). The lake basin has neither a prominent inlet nor a prominent outlet. It was probably formed by the melting of a buried ice block. The present lake is shallow, with a rather flat floor of organic sediment; maximum water depth is 4 m. Five borings were made along a transect across the lake, and the core chosen for study is near the center of the basin.

Rutz Lake is surrounded by farms, and no undisturbed stands of forest are close by. The Federal Land Survey records indicate the nature of the surrounding vegetation before clearance for agriculture, however. Table 2 gives witness-tree data for the township that contains Rutz Lake. It is clear that Rutz Lake was indeed situated in a deciduous forest rich in species. Elm and bur oak were the dominant trees, although both sugar maple and basswood were present.

TABLE 2. FOREST COMPOSITION IN T.116N., R.25W., AS CALCULATED FROM FEDERAL LAND SURVEY RECORDS OF 1855

Tree	Total no. of trees	Relative density	Relative dominance	Relative frequency	Importance value
Ulmus spp.	53	32.5	31.70	29.00	93.20
Quercus macrocarpa	25	15.4	18.00	14.10	47.50
Ostrya	20	12.0	6.55	14.10	32.65
Acer saccharum	9	5.4	13.30	6.50	25.20
Tilia	12	7.4	8.85	8.26	24.51
Quercus rubra +					
Q. ellipsoidalis	10	6.0	5.80	7.10	18.90
Populus tremuloides	11	6.8	3.46	6.50	16.76
Fraxinus	8	4.9	5.40	4.70	15.00
Carya	7	4.3	1.90	4.70	10.90
Quercus alba	2	1.2	4.35	1.60	7.15
Salix sp.	2	1.2	0.64	1.60	3.44
Juglans cinerea	2	1.2	0.50	1.60	3.30
Acer rubrum	1	0.6	0.44	0.78	1.82
Total	162				

The township that contains Rutz Lake was surveyed in September and October 1855. No signs of agricultural settlement were noted then, but the land was undoubtedly cleared and settled within a few decades after that date, and the mean date of forest clearance in the area is here assumed to be about 100 years ago.

METHODS

Sampling and Counting

Cores were taken with a modified Livingstone piston sampler of 2-inch diameter; below 11 m an A-frame and chain hoists were necessary to drive the sampler and a casing was used to support the drive rods (Cushing and Wright, 1965). The sediment was described in the laboratory according to the system of Troels-Smith (1955). It is essentially a slightly silty, marly copropel throughout; the important variations are noted on the pollen diagram (Pl. 1). Depths on the pollen diagram are measured from the water surface.

Volumetric samples were taken from the core with a small cylindrical corer that provided a sample of 0.5 cm^3 when pushed through a horizontal slice of core 9 mm thick. One sample from each slice of the core was used for pollen analysis, and a duplicate was dried and ignited at 500°C. The loss on ignition is shown on Figure 2 as a percent of dry weight.

Samples were prepared for pollen analysis by methods modified from those described by Faegri and Iversen (1964). The essential steps were: treatment with hot 10-percent KOH for 2 to 5 minutes, treatment with hot acetic acid, treatment with 48-percent HF for 15 minutes followed by dilution with 95-percent ethanol to reduce specific gravity before centrifugation, acetylation for 1 minute, staining with safranin, dehydration with tert-butanol, and mounting in silicone oil, 2000 centistokes.

Counting was done at a magnification of 300× with a fluorite objective (numerical aperture = 0.85). A slide area of 22 × 22 mm was scanned with nearly contiguous traverses, aided by a graticule with two parallel lines to limit the field of view to 280 μ. If an additional fraction of a slide was needed to bring the total count to more than 500 grains, the traverses were made in the central area of the cover glass.

A test of the reproducibility of the analysis was made on duplicate samples spaced 1 cm apart at the 880-cm level of the core. The pollen counts

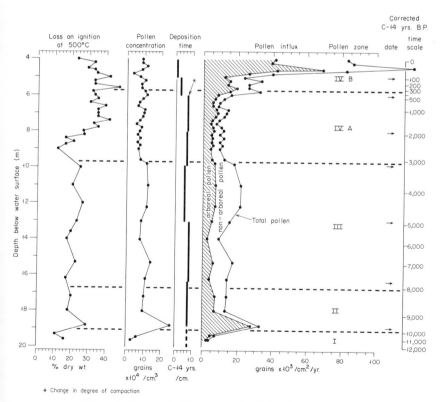

* Change in degree of compaction

Figure 2. Loss on ignition and pollen influx for Rutz Lake.

are shown in Table 3. A homogeneity test of these counts yielded a chi square of 40.6, with 23 degrees of freedom.

Identifications of pollen and spores were checked against the reference collection of the Limnological Research Center. All types determined in each sample are shown on the pollen diagram (Pl. 1) as a percent of the sum of all pollen and spores except those from obligate aquatic taxa and indeterminable grains. The categories of the latter follow Cushing (1964). A tabulation of the numbers of pollen grains and spores counted for each sample has been deposited with the National Auxiliary Publications Service.[1]

Determination of Pollen Concentration

The concentration of pollen and spores in each volumetric sample of sediment was determined by the method of addition of exotic pollen (Benninghoff, 1962). The exotic pollen used was *Ailanthus altissima*. A quantity of this pollen was washed and suspended in a solution of corn syrup adjusted to give a density approximately equal to that of the pollen grains. Phenol and thymol were added as preservatives. Withdrawals from the stock suspension were made while it was being mixed on a magnetic stirrer. Concentration of *Ailanthus* grains in the suspension was determined by counts of the pollen in a hemacytometer. Two concentrations were used in this study; one contained 8.4×10^4 grains/ml with a standard error of 0.12×10^4, the other 3.5×10^4 grains/ml.

One milliliter of the standardized suspension of *Ailanthus* pollen was added to each 0.5 cm³ of sediment before it was processed, and *Ailanthus* grains were counted together with fossil pollen and spores in the prepared slides. The number of fossil grains in the 0.5-cm³ sample of sediment could then be calculated by multiplying the number counted in the preparation by the factor S/A, where S is the number of *Ailanthus* grains added to the sample and A is the number of *Ailanthus* grains counted in the preparation. The values of pollen concentration are plotted as a curve on the pollen diagram (Pl. 1).

Pollen concentration in the duplicate samples at 880 cm was determined with the use of each of the two standardized suspensions of *Ailanthus* pollen. The two estimated values were 7.10×10^4 and 7.07×10^4 grains/cm³.

Determination of Charcoal Concentration

An estimate of the amount of charcoal in the sediment samples was based on a count of fragments encountered in the slides prepared for pollen analysis. Two fixed traverses were made of each slide, and all completely

---

[1] Order NAPS Document 00287 from the ASIS National Auxiliary Publications Service, c/o CCM Information Sciences, Inc., 22 West 34th Street, New York, New York 10001, remitting $1.00 for microfiche or $3.00 for photocopies.

TABLE 3. REPRODUCIBILITY OF
POLLEN DATA

	880 A	880 B
Pinus, subgen. *Haploxylon*	10	1
Pinus undiff.	44	22
Juniperus/Thuja	9	3
Alnus	4	10
Betula	9	10
Ostrya/Carpinus	13	8
Quercus	81	79
Ulmus	19	16
Fraxinus nigra	5	4
F. pennsylvanica	2	2
Populus	13	9
Salix	17	10
Acer	4	6
Other arboreal pollen	7	14
Gramineae	99	101
Cyperaceae	32	46
Artemisia	52	67
Ambrosia	40	48
Chenopodium-type	12	13
Urtica	6	3
Tubuliflorae	19	29
Other nonarboreal pollen	10	12
Aquatics	10	10
Unknown + indet.	5	0
Total:	522	523

charred fragments were measured. These consisted of uniformly opaque fragments of grass or sedge epidermis and woody tissue; partly charred tissue and chitin were ignored. The projected area of each fragment in the plane of the slide was estimated in comparison with a graticule grid with squares 15 μ on a side. Fragments were tallied by size classes; eleven classes were used, with mean area ranging from 0.5 to 27.5 grid squares, and larger fragments were measured and counted separately. Particles smaller than 1/4 grid square (56 μ^2) were ignored; they could not be distinguished readily from crystals of pyrite.

An estimate of the area of charcoal in the two traverses was obtained by summing, over all size classes, the product of the number of fragments in each class and the mean size for the class. This area was divided by the total area of the two traverses, and the quotient, expressed as percent, was multiplied by the same factor S/A used to calculate the number of pollen grains in the sample. The result is an index of the charcoal present in the sample; it is plotted as a curve on the pollen diagram (Pl. 1).

The charcoal index was determined independently for each of the duplicate samples at 880 cm. The values, 72 and 65, are considered to be in satisfactory agreement in comparison with the range of the index for all samples from the core.

RADIOCARBON DATES

Radiocarbon dates of eight samples from the core used for pollen analysis were provided by M. Stuiver, Yale Radiocarbon Laboratory. The dates are given in Table 4 and plotted against depth in Figure 3. All the samples dated were of copropel except the lowest (Y-1922), which was woody detritus from the base of the core.

The uppermost date (Y-1915) is of a sample just below an increase in pollen of *Ambrosia* (ragweed) in the pollen stratigraphy (Fig. 1). This increase,

TABLE 4. RADIOCARBON DATES FOR RUTZ LAKE

Depth below water surface (cm)	Depth below sediment surface (cm)	Sample no.	Date (uncorrected)	Date (corrected)
497–501	123–127	Y-1915	1100 ± 80	100 ± 80
598–602	224–228	Y-1916	1420 ± 100	420 ± 100
798–802	424–428	Y-1917	2920 ± 80	1920 ± 80
998–1002	624–628	Y-1918	4240 ± 100	3240 ± 100
1298–1302	924–928	Y-1919	5930 ± 100	4930 ± 100
1649–1653	1275–1279	Y-1920	8800 ± 160	7800 ± 160
1868–1872	1494–1498	Y-1921	10490 ± 120	9490 ± 120
1962–1970	1588–1596	Y-1922	12000 ± 160	

although not as pronounced as in other pollen diagrams from the area (McAndrews, 1969), is considered to mark the advent of land clearance by white settlers, taken to be 100 years ago in the area. The difference between this age and the radiocarbon date is 1000 years, and the discrepancy is attributed to the contribution to the lake of carbon from the leaching of limestone in the surrounding drift (Deevey and others, 1954; Ogden, 1967). Because the relative contribution from this source to the lake water undoubtedly varied with time, the 1000-year correction may not apply to all of the radiocarbon dates, but in the absence of evidence about the magnitude of such variation it seems reasonable to apply the correction uniformly to all the dates. A similar assumption was made by Ogden (1966) with a correction of 980 years and by Davis (1967) with a correction of 730 years. An exception is the lowest sample at Rutz Lake; because the sample dated was largely woody material of terrestrial origin, the correction should not apply.

The average rate of change of sediment age with depth is indicated in Figure 3 between each pair of radiocarbon dates. This rate, here called the deposition time, is also plotted in Figure 2, together with the total pollen concentration in grains per cubic centimeter. The pollen concentration of each sample divided by the deposition time in years per centimeter at the corresponding depth yields an estimate of the number of grains that were deposited at the site of the core on a unit surface area during a unit time. This quantity is here called the pollen influx; it is expressed as grains per square centimeter per year in Figure 2.

ZONATION AND INTERPRETATION OF THE POLLEN DIAGRAM

The diagram can be divided into four pollen zones that are thought to have regional significance. Nearly identical zones occur at Pickerel Lake,

South Dakota (Watts and Bright, 1968), and some of the zones can be identified in other diagrams from Minnesota, such as those of the Itasca transect (McAndrews, 1966).

Zone Rz I

In Rutz Lake this zone extends from the base of the lake sediment at 1968 cm to 1910 cm, dated respectively at 12,000 C^{14} years B.P. and 10,500 (corrected date = 9500) C^{14} years B.P. The zone is characterized by abundant pollen of *Picea, Populus, Alnus,* and *Larix.* This zone can rightfully be called the *Picea-Larix* pollen assemblage zone as described by Cushing (1967). It is almost identical with that of the basal zone I of Pickerel Lake. The trash layer of the basal sediment contains abundant needles, seeds, and cones of spruce. Aquatic pollen types include primarily *Potamogeton* (2 percent) and *Ruppia* (1 percent). The latter is reported for the first time from a Minnesota site.

The presence of a basal "trash" layer implies that Rutz Lake originated by the melting of an ice block buried by a superglacial forest. As the ice melted, the forest-floor debris collapsed into the newly formed lake (Florin and Wright, 1969). The water-logged soils of the initial depression were invaded by black spruce, tamarack, black ash, and ericaceous shrubs. Poplar,

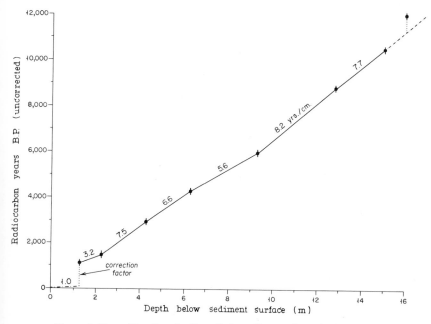

Figure 3. Deposition time for Rutz Lake sediments from the C^{14} dates.

birch, and alder probably grew in peripheral belts around the bogs and later around the ponds and lakes. Upland sites were probably occupied largely by white spruce and juniper. The driest, most exposed terrain may have featured prairie openings for Gramineae, and Compositae pollen total 10 to 30 percent. Kirchner Marsh and Lake Carlson (Wright and others, 1963) as well as Pickerel Lake also show nonarboreal pollen (NAP) values between 10 and 30 percent for this level. A boreal spruce forest is considered to have existed in these regions, as well as at Horseshoe Lake (Cushing, 1967), the Itasca area, and the Nebraska Sandhills (Watts and Wright, 1966). A full discussion of the late-glacial vegetation is given by Wright (1968).

Zone Rz II

This zone extends from 1910 to 1670 cm. A level near the upper zone boundary is dated as 8800 years B.P. (corrected date 7800 years B.P.). In Rutz Lake this zone is characterized by an increase of NAP from 20 percent to 60 percent. The arboreal pollen (AP) is represented principally by *Betula* and *Alnus* at the base and then by *Pinus, Quercus, Ulmus,* and *Ostrya/Carpinus.* The NAP consists primarily of Gramineae, Cyperaceae, *Artemisia, Ambrosia,* and *Chenopodium*-type. The aquatic pollen flora is dominated by *Potamogeton* and *Typha latifolia.*

In diagrams from farther northeast in Minnesota, such as those of Andree Bog (Cushing, 1964), Horseshoe and White Lily Lakes (Cushing, 1967), and Thompson Pond (McAndrews, 1966), this zone is called the *Pinus/Pteridium* zone or subzone, but those diagrams all have higher *Pinus* values and show a more pronounced peak of *Pteridium* spores than does the Rutz Lake diagram. The correlative zone at Pickerel Lake is zone 2.

Whereas at Pickerel Lake the pollen types that characterize the modern prairie pollen rain were not common in this assemblage zone (Watts and Bright, 1968), in Rutz Lake there are isolated grains of Leguminosae, *Potentilla, Mentha*-type, Cruciferae, and *Rumex,* as well as continuous occurrences of *Urtica, Iva ciliata,* and Tubuliflorae.

The pollen sequence implies an abrupt change in the vegetation as Rz I passes to Rz II. Within approximately 1000 years spruce was replaced by birch and alder and then by mesic deciduous trees. The relatively high values of *Quercus* (10 percent) and *Pinus* (16 percent) may represent grains of long-distance transport. Prairie openings increased during the time represented by the zone. These changes are considered to be the result of a change in climate toward warmer, drier conditions. Thus for sites outside the pine forest, such as Pickerel Lake and Rutz Lake, the vegetation of the time is believed to have consisted of a deciduous parkland with groves of trees on valley slopes or around depressions, where there was some protection from fire or where soil moisture was greater, perhaps in part because of accumulation of snow blown from the prairie (Watts and Bright, 1968). This type of vegetation resembles that described for the area at the time of settlement by Nicollet and by Winchell (Winchell, 1884b, p. 68–72). Similar vegetation extended north to Thompson and Terhell Ponds and east to the Kirchner Marsh area.

Zone Rz III

The percentages of individual arboreal pollen types drop to between 2 and 3 percent in this zone, except for oak and pine, which, as previously mentioned, are considered to have been blown in from a distance. The percentages of *Artemisia* and *Ambrosia* are higher than in the underlying zone. Figures for the major NAP types compare closely with the values obtained for correlative zones at other sites (Table 5). McAndrews (1966) called this zone at Terhell and Thompson Ponds the Gramineae-*Artemisia* assemblage zone.

In addition to the major components named above, several other prairie genera are present in minor quantities, such as *Petalostemum, Iva,* and *Psoralea.*

This zone extends from 1670 to 970 cm, encompassing the NAP maximum. It is dated as 8800 to 4240 B.P. (7800 to 3240 corrected).

The aquatic pollen flora includes *Potamogeton, Typha, Sparganium,* and *Myriophyllum,* which have several species of widely different environmental tolerances (Moyle, 1939). *Ruppia* has a maximum of 2 percent. This genus is known to require alkaline, hard-water lakes with silty bottoms.

Zone Rz III lasted approximately 4500 years. During this period prairie replaced the parkland, presumably as a result of the continued change to warmer, drier climate. The contemporaneous increase in the charcoal index suggests an increase in severity or frequency of fires. The charcoal for this interval is largely composed of grass or sedge epidermis, suggesting prairie fires rather than forest fires. Daubenmire (1936) and Curtis (1959) discuss

TABLE 5. MAJOR NONARBOREAL TAXA FOR RUTZ LAKE ZONE RZ III AND CORRELATIVE ZONES AT RELATED SITES*

Taxon	Rutz Lake	Terhell and Thompson Ponds	Pickerel Lake	Lake Carlson	Kirchner Marsh
Ambrosia-type	10–20	5–60	15–20	25	10–20
Artemisia	12–20	10–20	15–25	10	10–12
Tubuliflorae	4–6	5	4	<5	<5
Gramineae	20–25	15–30	15–20	15	20

* Figures in percent of total pollen.

at length the effects of fire on the prairie-forest boundary. Trees and shrubs bordering the prairie stand unsheltered in the path of the fire and are often severely damaged if not killed. The consequent reduction of shade may allow for further surface evaporation and drying of the vegetation cover, thus providing fuel for subsequent fires.

During this prairie phase the mesic deciduous woodland was restricted to groves with greater moisture or more fire protection, as it is in the western counties of Minnesota today and as described by Daubenmire (1936) for the western edge of the Big Woods. Many shallow lakes probably dried out, and the layer of compact sediment encountered in several small lakes in the region was formed. This layer may be correlated with the upper silt zone found in Iowa bogs (Walker, 1966). At Rutz Lake *Ruppia* pollen reaches a maximum during this interval; perhaps there was a greater accumulation of inorganic silts near the margin of the lake that would encourage the growth of this plant. No increase of inorganic sediment was noted in the main core at these levels.

The upland vegetation near Rutz Lake was probably prairie rather than oak savanna, because the percentage of oak pollen is not high. McAndrews (1966) found oak values of 10 to 17 percent in the surface cores from two ponds in the present prairie, and values of 15 to 25 percent in a core from a pond on the edge of the present oak savanna.

Zone Rz IV

A prominent rise in *Quercus* from 10 to 38 percent characterizes this zone, along with consistent occurrence of *Tilia*, *Acer saccharum*, *Corylus*, *Vitis*, *Platanus*, and *Tsuga*. The aquatic pollen flora gains *Nymphaea*, *Sagittaria*, *Brasenia*, and *Lemna*. The assemblage zone is divided into two subzones on the basis of relatively minor, but significant, changes in pollen composition.

Zone Rz IV A starts at 970 cm at the sharp rise in the AP curve, dated as 4240 B.P. (uncorrected). All the major NAP curves decrease by 1 to 6 percent, and all the major AP curves increase by 1 to 2 percent.

Zone Rz IV B starts at 570 cm at the increase of *Ulmus* from 1 percent to a maximum of 9 percent. Similarly, *Ostrya/Carpinus* increases from 1 to 7 percent and then decreases to an average value of 4 percent. *Pinus* decreases from 10 to 4 percent, and *Ambrosia* rises from 8 to 17 percent.

Zone Rz IV A is considered to represent gradual reforestation of the prairie as a result of a climate that was cooler and moister. A decrease in the severity or frequency of fires, indicated by a decrease in the charcoal index, would permit tree seedlings to become established. The vegetation could have been an open parkland, with groves of deciduous trees, mainly oaks and probably mainly *Quercus macrocarpa* (bur oak), growing in a prairie landscape of the type described by Winchell and by Curtis (1959) as oak openings. More deciduous species were present than at the time of the rather similar zone Rz II. The influx of the aquatic pollen flora (mainly rooted aquatics) may be attributed to shallowing of the lake as a result of sedimentation.

Zone Rz IV B represents the advent of the Big Woods in this area. Pollen of *Ulmus*, *Ostrya/Carpinus*, and *Tilia americana* reaches maximum values, and *Acer saccharum* is continuously present. These trees were common around Rutz Lake in 1855.

VARIATION IN POLLEN INFLUX

The number of pollen grains deposited per unit area per unit time is a measure of considerable interest at a site such as Rutz Lake, near the prairie-forest border. If the influx of pollen of a vegetation unit is sensitive to variation in the areal extent of that unit, as theory suggests, then shifts of the prairie-forest border ought to be recorded by stratigraphic changes in the influx of nonarboreal and arboreal pollen. The pollen influx at Rutz Lake does indeed show changes that are correlated with the pollen zones just described (Fig. 2).

Total pollen influx was lowest in zone Rz I and increased sharply in zone Rz II, from 3000 to 33,000 grains/cm²/year. Neither the pollen stratigraphy nor the spacing of radiocarbon dates has sufficient resolution to permit confident interpretation of this change, but it is consistent with the hypothesis that spruce forest, which disperses little pollen, was replaced by deciduous forest, with greater pollen productivity. The values of pollen influx in both zones also compare well with those reported by Davis (1967) from zones A and B in Connecticut, where the interpretation is quite similar.

The influx of nonarboreal pollen increased during deposition of zone Rz II to an average of about 11,000 grains/cm²/year in zone Rz III; it then decreased to about one-half that value in zone Rz IV B. This change supports the interpretation that prairie had its maximum extent during deposition of zone Rz III and was subsequently invaded by trees. The influx of arboreal pollen, however, fails to show the expected increase in zone Rz IV A.

In the uppermost zone, the pollen influx has much higher values, with a median of about 40,000 grains/cm²/year and a maximum 5 times greater than the average value of the lower zones. The increase involves both arboreal and nonarboreal pollen. An increase in arboreal pollen is consistent with the inferred forestation to produce the Big Woods; part of the nonarboreal pollen in zone IV B may have come from aquatic grasses, as suggested by the high and irregular percentages of Gramineae pollen. The magnitude of the increase, however, especially in comparison with the low values in zone IV A, remains an obstacle to the satisfactory explanation of the observed changes in pollen influx in terms solely of changes in vegetation.

An alternative hypothesis to explain the variation in pollen influx relates it to limnological events within the basin of Rutz Lake. Even if the pollen that impinges on a unit area of lake surface remained constant through time, changes in the pattern of deposition of the pollen and its matrix of sediment over the lake basin could result in apparent fluctuations in pollen influx at a single coring site in the basin. Thus if pollen and sediment were deposited evenly over the entire basin, the pollen influx to any unit area of sediment surface would be equal to the average influx to the water surface; but if currents in the basin caused a net movement of pollen and sediment from shallower to deeper parts of the basin, the pollen influx to the sediment in a deep part of the basin might be many times the influx to the water surface.

Evidence for such movement of pollen and sediment has been reported by Davis (1968) for a Michigan lake slightly smaller but deeper than Rutz Lake. There resuspension and mixing of sediment occurs throughout the lake basin, especially during spring and autumn overturn, and a net movement of sediment from shallow to deep water results in values for pollen influx in surface sediment that are 18 times greater in deep water near the center of the basin than in shallow water near the shore.

An important factor in determining the magnitude of sediment movement from shallow to deep water is the morphometry of the lake basin, particularly the proportion of the sediment surface that is above the thermocline and hence subject to currents and erosion of sediment during summer stratification. A second factor of uncertain importance is the density and distribution of rooted aquatic plants, which may reduce the net movement of sediment. Both of these factors may be sensitive to changes in lake level, which in turn may be climatically controlled.

There is evidence, reviewed by Wright (1968), from other lakes near the prairie-forest border that water levels dropped during the mid-postglacial expansion of the prairie and rose during the subsequent reforestation; to this may be added the observations already mentioned on sediment stratigraphy of lakes in the area of the Big Woods. It is conceivable, then, that the water level in Rutz Lake was relatively low during deposition of zone Rz III and that, as a result, a relatively large net transport of sediment and pollen occurred from shallower water to the deep water at the core site. An increase in the ratio of regional precipitation to evapotranspiration caused a rise in lake level concomitant with the change to parkland inferred from pollen zone Rz IV A; as the lake became deeper, newly deposited sediment was retained near the margins of the basin, and the pollen influx at the center decreased to values in zone IV A that may more nearly correspond to the influx at the lake surface. Continued deposition on sides of the basin, however, raised the sediment surface to the approximate level of the thermocline, and rapid transport of further increments of sediment to the center of the lake resulted both in the observed great increase in pollen influx in zone IV B and in the present profile of the basin floor, which slopes very gradually to the center. The increase in pollen of rooted aquatic plants from bottom to top of zone Rz IV supports the idea that the lake became progressively more shallow.

The curve for loss on ignition of the sediment (Fig. 2) is consistant with the hypothesis just presented. The sediment of zone Rz III has loss-on-ignition values of about 25 percent of dry weight, but in zone Rz IV A the values rise to exceed 30 percent. This increase in the organic fraction of the sediment suggests an increase in the ratio of autochthonous to allochthonous sediment and hence a decrease in the amount of mineral sediment from the shore to the site of the core (the $CaCO_3$ content of the sediment, as estimated by residual loss on ignition at 900° C, remained rather constant throughout the core at about 15 percent). The decrease in the organic fraction above 500 cm may be attributed to an increase in mineral sediment from erosion of the fields surrounding the lake after forest clearance.

A noticeable change in the degree of compaction of the sediment was noted at about 600 cm below the water surface, and the uppermost part of the sediment is highly flocculent. The calculated deposition time based on the radiocarbon dates for this part of the core is therefore quite low, and the average sedimentation rate before compaction is estimated to be as high as 1 cm/year. Rutz Lake receives runoff from four farms, and its productivity is undoubtedly high.

Probably neither of the hypotheses presented here accounts alone for all of the observed variation in pollen influx in the core from Rutz Lake, and the pollen influx was affected both by changes in the upland vegetation and by changes in lake level and patterns of deposition within the lake basin. Both of these factors, however, are coupled to climatic change, and so the correspondence between change in pollen influx and the pollen zonation is understandable even if the direction and magnitude of the change remains poorly understood.

CONCLUSIONS

Although interpretation of pollen influx at Rutz Lake is complicated by processes occurring within the lake, the same processes of erosion, mixing, and redeposition of sediment help ensure that the pollen percentages in a single core are a reliable average of the composition of the pollen entering the lake basin (Davis, 1968). Thus the pollen stratigraphy based on relative frequencies is consistent with the pollen stratigraphy at other sites in Minnesota and South Dakota and supports the regional history outlined by Wright

(1968) for the northeastern edge of the Prairie Peninsula. The variation of the charcoal index at Rutz Lake testifies to the importance of fire in the prairie environment and to the inverse relation between the prevalence of fire and the degree of tree cover near the prairie-forest border. Although the frequency of wildfire may be directly controlled by climate, a prairie fire typically covers many miles until arrested by a fire break (*see* historical accounts in Curtis, 1959). The sequence at Rutz Lake suggests that increased soil moisture allowed the expansion of trees from areas protected from fire by natural fire breaks onto more exposed sites and that the expansion and coalescence of groves of trees provided an obstacle to fires that continued to sweep the prairie to the west and south.

The stratigraphy at Rutz Lake suggests a rather recent origin for the Big Woods in the surrounding area, as little as 400 years ago, and lends support to the suggestion by McAndrews (1969) that succession toward a stable maple-basswood forest has proceeded up to the present, perhaps quickened by the even greater reduction in fire since white man settled the area about 100 years ago.

REFERENCES CITED

Benninghoff, W. S., 1962, Calculation of pollen and spore density in sediments by addition of exotic pollen in known quantities (abstract): Pollen et Spores, v. 4, p. 332–333.

Bray, J. R., 1956, Gap phase replacement in a maple-basswood forest: Ecology, v. 37, p. 598–600.

Curtis, J. T., 1959, The vegetation of Wisconsin: Madison, Wisconsin Univ. Press, 667 p.

Cushing, E. J., 1964, Redeposited pollen in late-Wisconsin pollen spectra from east-central Minnesota: Am. Jour. Sci., v. 262, p. 1075–1088.

—— 1967, Late-Wisconsin pollen stratigraphy and the glacial sequence in Minnesota, p. 59–88 in Cushing, E. J., and Wright, H. E., Jr., *Editors*, Quaternary paleoecology: New Haven, Connecticut, Yale Univ. Press, 433 p.

Cushing, E. J., and Wright, H. E., Jr., 1965, Hand-operated piston corers for lake sediments: Ecology, v. 48, p. 380–384.

Daubenmire, R. F., 1936, The Big Woods of Minnesota: its structure and relation to climate, fire and soils: Ecol. Mon., v. 6, p. 233–268.

Davis, M. B., 1967, Pollen accumulation rates at Rogers Lake, Connecticut, during late- and postglacial time: Rev. Palaeobotany Palynology, v. 2, p. 219–230.

—— 1968, Pollen grains in lake sediments: redeposition caused by seasonal water circulation: Science, v. 162, p. 796–799.

Deevey, E. S., Jr., Gross, M. S., Hutchinson, G. E., and Kraybill, H. L., 1954, The natural C^{14} contents of materials from hard-water lakes: Natl. Acad. Sci. Proc., v. 40, p. 284–288.

Faegri, Knud, and Iversen, Johs., 1964, Textbook of pollen analysis: New York, Hafner Publishing Co., 2nd edition, 237 p.

Fernald, M. L., 1950, Gray's manual of botany: New York, American Book Co., 8th edition, 1632 p.

Florin, M. B., and Wright, H. E., Jr., 1969, Diatom evidence for the persistence of stagnant glacial ice in Minnesota: Geol. Soc. America Bull., in press.

Janssen, C. R., 1966, Recent pollen spectra from the deciduous and coniferous-deciduous forests of northeastern Minnesota: A study in pollen dispersal: Ecology, v. 47, p. 804–825.

—— 1967, A comparison between the recent regional pollen rain and the sub-recent vegetation in four major vegetation types in Minnesota (U.S.A.): Rev. Palaeobotany Palynology, v. 2, p. 331–342.

McAndrews, J. H., 1966, Post-glacial history of prairie, savanna, and forest in northwestern Minnesota: Torrey Bot. Club Mem., v. 22, no. 2, 72 p.

—— 1969, Pollen evidence for the protohistoric development of the "Big Woods" in Minnesota, U.S.A.: Rev. Palaeobotany Palynology, in press.

Moyle, J. B., 1939, The larger aquatic plants of Minnesota and the factors determining their distribution: Minneapolis, Minnesota Univ., Ph.D. dissert.

Nicollet, J. N., 1841, Report intended to illustrate a map of the hydrogeographical basin of the upper Mississippi River: Senate Document 26th Congress 2nd session, v. 237 (Serial Set v. 380).

Ogden, J. G., III, 1966, Forest history of Ohio. I. Radiocarbon dates and pollen stratigraphy of Silver Lake, Logan Co., Ohio: Ohio Jour. Sci., v. 66, p. 387–400.

—— 1967, Radiocarbon and pollen evidence for a sudden change in climate in the Great Lakes region approximately 10,000 years ago, p. 117–130 in Cushing, E. J., and Wright, H. E., Jr., Editors, Quaternary paleoecology: New Haven, Connecticut, Yale Univ. Press, 433 p.

Troels-Smith, J., 1955, Characterization of unconsolidated sediments: Danmarks Geol. Undersøgelse, IV Raekke, v. 3, no. 10, p. 1–87.

Walker, P. H., 1966, Post-glacial environments in relation to landscape and soils on the Cary drift, Iowa: Agriculture and Home Economics Experiment Station, Iowa State Univ. Agr. Expt. Sta. Res. Bull. 549, p. 839–875.

Watts, W. A., and Bright, R. C., 1968, Pollen, seed, and mollusk analysis of a sediment core from Pickerel Lake, northeastern South Dakota: Geol. Soc. America Bull., v. 79, p. 855–876.

Watts, W. A., and Wright, H. E., Jr., 1966, Late-Wisconsin pollen and seed analysis from the Nebraska Sandhills: Ecology, v. 47, p. 201–210.

Winchell, N. H., 1884a, The Geological and Natural History Survey of Minnesota: 12th Annual Report.

—— 1884b, The geology of Minnesota: Minnesota Geol. Nat. History Survey Final Rept., v. 1, 697 p.

—— 1888, The geology of Minnesota: Minnesota Geol. Nat. History Survey Final Rept., v. 2, 695 p.

Wright, H. E., Jr., 1968, History of the prairie peninsula, p. 78–88 *in* Bergstrom, R. E., *Editor*, The Quaternary of Illinois: Illinois Univ. Agr. Spec. Pub. 14, 179 p.

Wright, H. E., Jr., and Ruhe, R. V., 1965, Glaciation of Minnesota and Iowa, p. 29–42 *in* Wright, H. E., Jr., and Frey, D. G., *Editors*, The Quaternary of the United States: Princeton, New Jersey, Princeton Univ. Press, 922 p.

Wright, H. E., Jr., Winter, T. C., and Patten, H. L., 1963, Two pollen diagrams from southeastern Minnesota: problems in the regional late-glacial and post-glacial vegetational history: Geol. Soc. America Bull., v. 74, p. 1371–1396.

CONTRIBUTION NO. 73, LIMNOLOGICAL RESEARCH CENTER, UNIVERSITY OF MINNESOTA

GEOLOGICAL SOCIETY OF AMERICA, INC.
SPECIAL PAPER 123

Vegetational History of the Shenandoah Valley, Virginia

ALAN J. CRAIG

Limnological Research Center, University of Minnesota,
Minneapolis, Minnesota .

ABSTRACT

Two diagrams show the results of pollen analysis in sediments from Hack and Quarles Ponds, in the southern part of the Shenandoah Valley, Virginia. The following pollen assemblage zones are distinguished:

(1) A basal *Pinus-Picea* zone, with abundant conifer pollen and relatively large amounts of herb pollen. Two subzones are distinguished: a *Sanguisorba-Isoetes* subzone, present only at Hack Pond, where the top of the subzone has been dated at 12,720 ± 200 years, and a *Quercus-Corylus* subzone, with less herb and more deciduous-tree pollen. This zone is believed to represent a period of conifer woodland dominated by spruce.

(2) A *Quercus* zone, with abundant *Quercus* and very little conifer pollen. Two subzones are distinguished: a lower *Tsuga* subzone and an upper *Carya-Cephalanthus* subzone. This zone is believed to represent a period when the vegetation was largely hardwood forest, commencing shortly after 9520 ± 200 years ago.

(3) A *Quercus-Pinus* zone, with much pollen of *Quercus* and *Pinus* and substantial amounts of other deciduous-tree pollen. This zone is believed to represent a period of oak-pine forest, the pines being predominantly those species found today in the vicinity of the sites.

The diagrams strongly resemble those from southeastern Virginia and North Carolina, but they differ very markedly from those of sites to the north in the area formerly covered by ice sheets of the Wisconsin glaciation. A time lapse probably occurred between similar vegetational developments in different areas. The results support the conclusions of some previous writers that substantial changes in vegetation occurred at considerable distances from the ice sheets. There is also support for suggestions that the Appalachians constituted a refuge for *Picea* and *Pinus* during at least part of Wisconsin time.

CONTENTS

INTRODUCTION

Numerous studies have been made of postglacial pollen successions in eastern North America north of the border of Wisconsin glacial drift, but little information is available for south of the drift border. Our knowledge of the subject was reviewed by Whitehead (1965). Subsequent studies were reported by Cox (1968) and Watts (1969, unpublished manuscript).

The present study concerns two profiles from sites in the southern Shenandoah Valley in Virginia. This part of the Appalachian belt, lying about 200 miles south of the glacial limit, is an important location phytogeographically. It may have constituted a refugium from which plant species could have spread northward as the ice sheet retreated. The biogeographic context is very different from that in western Europe, in which the Alps provided a barrier to north-south migration of plant species.

ACKNOWLEDGMENTS

The major part of this work was carried out at the University of Dublin. I should like to thank W. A. Watts for his initial instruction in pollen analysis and for sustained advice and discussion. The work was completed at the Limnological Research Center, University of Minnesota, and I should like to thank H. E. Wright, Jr., who critically read the manuscript and provided invaluable guidance and assistance in its final preparation. Thanks are also due to J. T. Hack for suggesting the sites and supplying geological data, to W. A. Watts, H. E. Wright, Jr., and R. O. Megard for obtaining the core and providing ecological information on the region, to R. Keatinge and O. Anderson for technical advice and assistance, and to M. Stuiver, Yale University, for radiocarbon dating. The work was supported financially by National Science Foundation Grants GB-3401 and GB-7163 to the University of Minnesota.

SITES AND STRATIGRAPHY

The cores studied were obtained from two ponds; one will be referred to as Hack Pond; the other, named Lake Quarles on some maps, will be referred to as Quarles Pond. Hack Pond lies at latitude 37°59′5″N., longitude 78°59′50″W., at an elevation of 1540 feet, 2 miles southwest of Sherando, Augusta County, Virginia, in the Lovingston quadrangle of the U.S. Geological Survey. Quarles Pond lies at latitude 37°59′45″N., longitude 79°4′20″W., at an elevation of 1640 feet, about 4 miles west of Hack Pond in the Vesuvius quadrangle.

The area is underlain by Early Cambrian dolomite in which sinkholes are partially filled with residual soil and alluvium washed down from the Blue Ridge to the southeast. Hollows have subsequently developed in the alluvium over the sinkholes (Hack, 1965). Hack and Quarles Ponds are two such hollows, lying in the basin of the South River, a tributary of the Shenandoah River, at the foot of the northwestern slope of the Blue Ridge.

Hack Pond contains approximately 2 m of sediment beneath 30 cm of water. The core was described in the field (Table 1). Subsequently, in the laboratory, values for percent loss of weight on ignition at 650° C were calculated as a measure of the organic content of the sediments (Pl. 1). The untreated sediment is rich in diatoms, especially near the top of the core. The pH of the water on November 7, 1965, was 5.7, the hardness 0.11 milleequivalents per liter, alkalinity 0.09 me/L, Ca^{++} 0.06 me/L, and Mg^{++} 0.05 me/L.

Quarles Pond is also shallow; a core of sediment 177 cm long was obtained in water only 15 cm deep (Table 2).

TABLE 1. SEDIMENT STRATIGRAPHY OF
HACK POND

Depth (cm)	
30–50	Dark-brown copropel with roots
50–60	Dark-brown clayey copropel with roots
60–80	Dark-brown clayey copropel
80–92	Light-brown copropelic clay
100–123	Weakly banded gray and brown silty clay
123–148	Gray silty clay with vertical black streaks
150–181	Light-brown sandy silt
181–211	Gray fine sandy silt with a few black flecks

TABLE 2. SEDIMENT STRATIGRAPHY OF
QUARLES POND

Depth (cm)	
15–50	Dark-brown copropel with roots
50–65	Dark-gray copropelic clay
65–96	Dark-gray clay with bands of organic flecks
100–125	Dark-gray clay
125–140	Light-gray clay
140–150	Gray silty sand
150–192	Gray to brown silty sand

The following trees and shrubs were observed by W. A. Watts in the present vegetation surrounding Hack Pond. *Quercus alba, Q. rubra, Q.* cf. *velutina, Q. marilandica, Q. prinus, Pinus virginiana, P. pungens, P. rigida, Carya cordiformis, Castanea dentata* (diseased shoots), *Juniperus virginiana,* cf. *Carpinus caroliniana, Vaccinium* sp., *Rhododendron (Azalea)* sp., *Kalmia latifolia, Lyonia ligustrina, Gaultheria procumbens, Epigaea repens,* and *Smilax glauca.* The site lies within the oak-chestnut forest region of Braun (1950). However, the vegetational map prepared by Küchler (1964) shows a transition from Appalachian oak forest (formerly oak-chestnut forest) to oak-hickory-pine forest close to the site. The list of species above is suggestive of such a transition.

The location of the two sites and their relation to the Wisconsin drift border and to present natural vegetation regions are shown in Figure 1. Also shown are locations of other sites referred to in the following discussion.

METHODS

Sample Preparation and Examination

Pollen samples were prepared for examination by standard methods (Faegri and Iversen, 1964). Treatment with HCl was not necessary. Slides were mounted in silicone oil.

All routine counts were made with a Leitz Ortholux microscope. Identifications were made by comparison with slides of modern pollen and with photomicrographs (Erdtman and others, 1961). Most of the samples contained abundant pollen, and a count of 300 grains was completed in almost every case after examination of less than one whole slide.

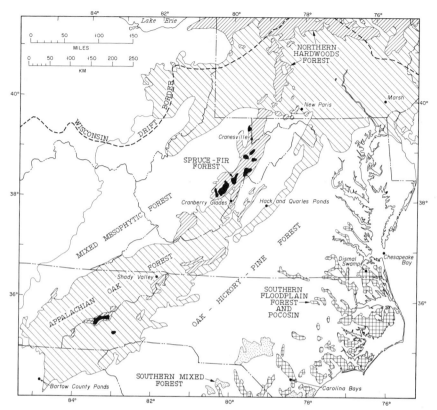

Figure 1. Vegetation map showing location of sites mentioned in the text. Vegetation boundaries *after* Küchler (1964).

Tabulation

The pollen sum consisted initially of 300 grains, excluding floating and submersed aquatics. Such a pollen sum is thought to be the most objective in an investigation of regional pollen succession (Wright and Patten, 1963). At the same time, counts were made of pollen of aquatic plants and of spores of Pteridophyta and *Sphagnum*. Subsequently, pollen of *Cyperaceae* and *Cephalanthus* were also excluded from the pollen sum. These grains were abundant at Quarles Pond at some levels and to a lesser extent at Hack Pond; they are believed to be almost entirely of local origin. As such they would distort the regional pollen rain, which was the object of the study. All types recognized are included in the diagrams. Unknowns, which rarely exceeded 2 percent of grains counted, are not included.

Some names of taxa are qualified, and the qualifications require explanation. The prefix *cf.* signifies close agreement between the fossil grains of the taxon following the prefix, but it indicates a lack of sufficient reference material to permit exclusion of other taxa from consideration. The suffix -*type* indicates that several taxa are included as possibilities: *Dryopteris*-type includes all monolete spores without a preserved perispore; *Sparganium*-type includes *Typha angustifolia* as well as all species of *Sparganium*.

Zonation of the Diagrams

The pollen diagrams (Pl. 1) were divided into pollen assemblage zones and subzones, in accordance with the Code of Stratigraphic Nomenclature (American Commission on Stratigraphic Nomenclature, 1961). These zones are biostratigraphic units and more specifically are fossil assemblage zones (Article 21 of the Code). They carry no implications concerning the time of deposition or the climate and vegetation at that time. In view of the similarity of the two diagrams and the geographic closeness of the two sites, it was possible to use the same zonation for both, although the lowest subzone at Hack Pond is absent from Quarles Pond.

Additional Investigations

After routine pollen analyses had been completed, slides of some samples were re-examined in an attempt to determine the contributions of different species of *Pinus*. The pollen of *Pinus strobus* can be distinguished from that of other species of *Pinus* occurring at present in the eastern United States by the presence of verrucae on the furrow between the bladders. However, only well-preserved grains can be assigned to one type or the other. The ratios of the two types are shown on the *Pinus* curve in the Hack Pond diagram. The number of determinable grains on which the ratio is based is also shown for each of the 13 samples examined. It should be noted that some of these numbers are extremely low.

A study was also made of seeds and other plant macrofossils present in the core from Hack Pond. Although the seed flora was too meager to justify quantitative presentation, some identifications that cast further light on the pollen diagram are mentioned in the descriptions of the zones.

THE POLLEN ZONES AND THEIR VEGETATIONAL SIGNIFICANCE

Pinus-Picea Zone

This zone is characterized by 40 to 60 percent *Pinus* pollen, which at Hack declines abruptly at the upper boundary. *Picea* is the second major constituent, showing fluctuating curves with a peak of 39 percent at Hack. *Abies* and *Ostrya/Carpinus* are present in small quantities and are almost confined to this zone. Nonarboreal pollen types are generally more common in this zone than elsewhere, for example, *Tubuliflorae*, *Sanguisorba*, and *Umbelliferae*. The spores of pteridophytes, particularly *Lycopodium*, occur occasionally. *Isoetes* and *Myriophyllum* are both persistent throughout the zone at Hack and are almost confined to it, and *Nuphar* is almost confined to the middle of the zone. At Quarles, however, pollen of aquatic plants is scarce in this zone.

    Sanguisorba-Isoetis Subzone. *Sanguisorba* is almost confined to this sub-zone, declining gradually from 5 percent at the base. *Isoetes* is very abundant, attaining a peak of 97 percent. This peak seems to be due primarily to *I. echinospora;* another species, probably *I. riparia*, was also present. *Picea* is persistently high, and pollen of deciduous trees is scarce. There are no spectra at Quarles that fit this subzone. The most likely explanation is that deposition started later at Quarles than at Hack. It could be argued that *Sanguisorba* pollen and *Isoetes* macrospores of the basal levels at Hack were of local origin and that the basal sediments at the two sites are actually of the same age. However, the high percentages of *Quercus* and other deciduous-tree pollen in the lowest spectra at Quarles would seem to rule out such a hypothesis and also to rule out the possibility of mixing of sediments.

    Quercus-Corylus Subzone. *Quercus* and *Corylus* are more common than below; at Quarles *Quercus* is a major constituent. Only in this subzone and at the top of the lower subzone does the proportion of *Pinus* grains of *strobus* type exceed 10 percent. *Picea* and *Isoetes* are less common than below. *Ericales* pollen is persistent in the upper part of the zone.

    The *Pinus-Picea* pollen assemblage zone represents some form of boreal woodland. *Pinus* is overrepresented in pollen samples in eastern North America, and *Picea* and *Abies* are at most proportionately represented (Carroll, 1943; Davis and Goodlett, 1960). *Betula* and *Quercus* are usually overrepresented; consequently, their presence in the diagrams at this level may not be significant. *Picea* was probably the dominant tree, with some *Pinus* and *Abies* and a few deciduous trees. The principal *Pinus* species may have been *P. banksiana*, according to pollen and needle evidence from northern Georgia (Watts, 1969, unpublished manuscript). The *Pinus* pollen ratios at 175 and 160 cm indicate that *P. strobus* was present at that time but was not a major component of the vegetation. The ratios for succeeding samples are based on very few grains and may not be significant. Non-arboreal pollen (NAP) exceeds 15 percent only in the lowest sample at Hack. Although no quantitative data are available on pollen representation of herbs, only very much larger percentages of NAP have been taken by other writers to indicate predominance of open vegetation. Davis and Goodlett (1960), however, found NAP values of 6 to 16 percent where 45 percent of the landscape was unforested. Therefore, considerable open spaces probably existed. Some of these could have taken the form of boggy lake margins. *Sanguisorba*, for example, is a plant of boggy soils today (Whitehead, 1965).

Quercus Zone

The base of this zone is marked by a rise of *Quercus* and fall of *Pinus*, sharp at Hack and gradual at Quarles. *Picea* disappears gradually, but pollen of *Tsuga* and numerous deciduous trees becomes important. Of herbs, *Tubuliflorae*, *Ambrosiae*, and *Gramineae* are present at both sites in small quantity, and *Chenopodiaceae/Amaranthaceae* is important at Quarles. Outside the pollen sum, *Cephalanthus* is important, especially at Quarles. *Sphagnum* and several pteridophytes occur. *Brasenia* is the most important aquatic type at Hack, and *Sparganium*-type at Quarles.

The contrasts between the two sites in the curves for pollen types of presumably local origin, for example, *Sparganium*, *Sagittaria*, and *Brasenia*, probably reflect differences in the nutrient status of the two ponds. It was noted in the field (W. A. Watts, oral commun.) that Quarles Pond seems to be more eutrophic than Hack Pond at the present day. Differences in vegetation in and around the ponds may also account for the greater importance of *Cephalanthus* at Quarles and of *Alnus* at Hack.

Tsuga **Subzone.** *Tsuga* shows a distinct peak in this subzone. *Picea* almost disappears at the top of the subzone. *Castanea* appears for the first time. Pteridophytes are relatively common, but *Cephalanthus* is sparse. Chenopod pollen attains a maximum at Quarles.

Carya-Cephalanthus Subzone. *Carya* and *Cephalanthus* are more common than below, as is *Quercus*. *Pinus* reaches its over-all minimum, and *Tsuga* declines.

The *Quercus* pollen assemblage zone (particularly the lower subzone) may represent vegetation resembling the northern hardwood forests today. *Pinus* was probably very minor in the vegetation, and *Quercus* pollen may be heavily overrepresented. *Acer* was found to be underrepresented by Davis and Goodlett (1960). Little information is available on representation of the other forest species involved. NAP values at Hack range from 1 to 11 percent; some higher values occur at Quarles due primarily to a peak of chenopods. Some of the higher figures may signify considerable open vegetation. However, it must be borne in mind that some of the pollen included as NAP could come from plants of wet situations. This applies particularly to the *Tubuliflorae*.

Quercus-Pinus Zone

Quercus is the most common type, although less abundant than in the zone below. *Pinus* increases at the base and continues to rise slowly. Some pine-needle fragments were identified by W. A. Watts as *Pinus rigida*, which is found in the local vegetation today, indicating that this species was one of those contributing pollen. *Carya, Castanea,* and *Betula* are fairly constant, although never abundant, and are accompanied by other deciduous tree types. Herb pollen is sparse, although *Gramineae* is always present. *Cyperaceae, Brasenia,* and *Potamogeton* occur at both sites, and *Sagittaria* is also present at Quarles. Seeds of *Cyperaceae* and *Potamogeton* are common. Of these, seeds of *Psilocarya scirpoides* are most abundant, followed by *Potamogeton spirillus.*

It is probable that several vegetation types contributed to the pollen rain in the *Quercus-Pinus* zone, which extends to the surface of the sediments. Such a situation may also be represented by earlier zones. Certainly it would appear that *Quercus-Castanea* forests, *Quercus-Pinus-Carya* forests, and mixed-hardwood forests have all contributed in recent years, and the pollen assemblage zone is fairly uniform. Again the available evidence suggests that *Quercus, Pinus,* and *Betula* are overrepresented, and *Acer* and *Tilia* underrepresented. NAP values never exceed 4 percent except at the base of the zone and at the surface at Quarles. This suggests no extensive open vegetation during deposition of this zone. It is noteworthy that there is a reasonable correspondence between the present vegetation at Hack Pond, described earlier, and the uppermost spectra of the two diagrams.

POSSIBLE CLIMATIC CONDITIONS

Many of the genera present in the *Pinus-Picea* zone are somewhat boreal in their distribution, notably *Picea, Abies,* and *Sanguisorba.* This would suggest a climate more or less similar to that of boreal regions today. *Polemonium,* on the other hand, has a decidedly temperate distribution, as do *Quercus* and *Ostrya* (or *Carpinus*) (Davis, 1958). It is thus clear that the climate cannot have been extreme, although the possibility of long-distance transport of some of the pollen grains of temperate plants cannot be ruled out.

During deposition of the *Quercus* and *Quercus-Pinus* zones the climate cannot have been drastically different from that at present. The *Quercus* zone probably records a period intermediate in warmth between the other two zones, when conditions resembled those in northern hardwood forests at present. There is no evidence for climatic change since the beginning of the *Quercus-Pinus* zone, nor is there any evidence in the diagrams of a hypsithermal interval. The present climate of the region is moderately dry, and it is quite possible that past changes of vegetation reflect changes in water availability. *Cephalanthus* is a swamp shrub today (Davis, 1958). The maxima of *Cephalanthus* and *Sphagnum* may or may not indicate such changes; they may reflect local conditions only.

COMPARISON WITH POLLEN SUCCESSION AT OTHER SITES

In Maine, well north of the drift border, five pollen zones and subzones were recognized by Deevey (1951). The diagrams are dominated successively by pollen of (A) *Picea* and *Abies*, (B) *Pinus*, (C-1) *Quercus* and *Tsuga*, (C-2) *Quercus* and *Carya*, and (C-3) *Quercus* and *Castanea*. Variants of this pattern are found in other parts of New England (Davis, 1958, 1960; Ogden, 1963; Whitehead and Bentley, 1963). Even much farther west comparable patterns were recognized by Sears in the Great Lakes region (Ogden, 1965). Even in Pennsylvania just south of the drift border a similar sequence was observed by Martin (1958), although the lowest zone there was dominated by herbs rather than *Picea*. Guilday and others (1964) presented a very different diagram from a cave deposit, with *Pinus* predominant throughout, but there are possibilities of stratigraphic mixing, so the significance is uncertain.

It is not surprising to find only a slight resemblance between the Virginia diagrams and those from close to or north of the drift border, for ecological conditions must have been very different in Virginia from what prevailed in New England and the Great Lakes region immediately after glaciation. Furthermore, most of Virginia today has *Quercus-Pinus-Carya* forest as its natural vegetation, whereas *Quercus* (formerly *Quercus-Castanea*) forest is predominant in Pennsylvania and northern hardwoods forest in New England. It might be thought that the general pollen succession would be the same, with perhaps an extra zone at the base in the north and at the top in the south, each pollen zone being diachronous. Such a simple situation is not found. *Pinus*, rather than remaining important and dominating another zone after the decline of *Picea*, actually declines at about the same level as does *Picea*. The possible significance of this will be discussed later. For subsequent zones, there is some resemblance between the successions in Virginia and New England. In Virginia a *Tsuga* maximum does occur before a *Carya* maximum, but no distinct *Castanea* maximum occurs. The Virginia diagrams bear more resemblance to that of Martin (1958) from Pennsylvania than to diagrams from farther north, with respect to relative abundance of *Carya*, *Tsuga*, *Fagus*, and so forth.

Four investigations in the Appalachian belt have been reported. The results presented by Cox (1968) from a site on the West Virginia–Maryland border show greater similarities to the present work than do those already discussed. The *Picea* and *Pinus* peaks in zones 1 and 2 and the *Carya* and *Castanea* peaks in zone C-2 correlate well with features in the Hack and Quarles diagrams. The higher values of *Picea* and NAP in the lower zones and of *Tsuga* and *Fagus* in the upper, as well as the return of *Picea* at the top, are probably related to the more northerly position of the site and the greater elevation (2500 feet).

The nearest investigation to the sites under discussion was by Darlington (1943) at Cranberry Glades, West Virginia. Very few pollen types were recognized, however. Successive zones were dominated by (1) *Picea* and *Abies*, (2) *Pinus* and *Betula*, (3) *Tsuga*, *Betula*, and *Quercus*, and (4) *Quercus*, *Betula*, *Pinus*, and *Carya*. This sequence bears little more resemblance to my results than do sequences from New England. The unpublished diagram from Shady Valley, Tennessee, constructed by Barclay and described by Whitehead (1965), bears even less resemblance to those under discussion, particularly in the presence of much *Tsuga* and the increase of *Picea* and *Abies* at the top. However, the diagram is from a high-level bog where *Picea* grows at present, and consequently it may not be representative of regional conditions.

Watts (1969, unpublished manuscript) obtained results from Bartow County, northwestern Georgia, that can be clearly correlated with those from Hack and Quarles Ponds. The upper part of zone Q1 and all of zone Q2 correspond to the *Pinus-Picea* zone, zone Q3 to the *Quercus* zone, and zone Q4 to the *Quercus-Pinus* zone. The lesser amounts of *Picea* and some hardwoods and the greater importance of *Pinus* must be due to the much more southerly location of the sites. Watts formally describes a *Pinus-Picea* zone at sites in Georgia and North Carolina, and he defines the criteria by which it may be recognized. The zone of the same name at Hack and Quarles Ponds does not fall within the limits of his definition, which might therefore be extended to include these sites.

Numerous diagrams exist from the coast of Virginia and the Carolinas. A summary diagram for the Dismal Swamp, Virginia (Whitehead, 1965), shows a zone 1 that resembles the upper part of the *Pinus-Picea* zone at Hack and Quarles Ponds, although *Picea* is less common and *Carya* more persistent. Herbs such as *Thalictrum* and *Sanguisorba* seem to have persisted later relative to other types. Zone 2 resembles the *Tsuga* subzone, particularly in the maxima of those two genera. Zone 4 does not clearly resemble the *Quercus-Pinus* zone, as it is dominated by pollen of *Taxodium*, which must have grown locally.

Diagrams from boreholes at the mouth of Chesapeake Bay (Harrison and others, 1965) show a *Pinus-Picea* zone that strongly resembles the *Pinus-Picea* zone at Hack and Quarles, although with more NAP despite a lack of *Sanguisorba*. This is succeeded by a *Tsuga-Betula-Acer-Fagus* zone that probably corresponds to the *Tsuga* subzone, and then by a *Quercus-Carya* zone resembling the *Carya-Cephalanthus* subzone. The diagrams are truncated above.

A generalized diagram is also available for part of North Carolina (Frey, 1953). The lower parts of the diagram are not relevant here. The upper part bears a marked resemblance to those from Hack and Quarles. The M3 zone is very similar to the *Pinus-Picea* zone and even shows a high

Isoetes peak such as occurs in the lowest subzone at Hack. The quantity of *Picea* is much less, as at Dismal Swamp. The Ca zone is similar to the *Quercus* zone, although not as clearly differentiated. The minimum of *Pinus*, disappearance of *Picea*, and maxima of *Quercus*, *Carya*, and *Tsuga* are all present. The Cb zone is like the *Quercus-Pinus* zone, although the greater amounts of *Pinus* and *Nyssa* and the presence of *Liquidambar* and *Taxodium* indicate regional variation, easily comprehensible in terms of present differences in vegetation.

The diagrams from Hack and Quarles Ponds thus have strong affinities with those from coastal Virginia and the Carolinas and to a lesser extent with those of Cox (1968) from the higher mountains of Maryland and of Watts (1969, unpublished manuscript) from the piedmont of northern Georgia. Although similarities occur, there is no proof of synchroneity between similar developments at different sites. In fact, a northward progression of gross vegetation types is suggested. Specifically, the top of the *Pinus-Picea* zone is dated as older than $10,224 \pm 510$ in southeastern North Carolina (Whitehead, 1965), whereas at Hack Pond it slightly postdates 9520 ± 200 (Y-2594), and at nearby Cranberry Glades it is dated as 9434 ± 840 (Guilday and others, 1964). A similar time lapse between events in northern Georgia and northeastern Virginia is indicated by dates for Quicksand Pond (Watts, 1969) and Hack Pond. At Quicksand the base of zone Q2, that is, the point at which the rise of *Quercus* and fall of *Pinus* begin, has been dated to $13,560 \pm 180$ years B.P. At Hack Pond, on the other hand, the *Quercus* curve begins to rise only after a date of $12,720 \pm 200$ (Y-1980) at the top of the *Sanguisorba-Isoetes* subzone, and attains very high values only much later.

On the basis of the established dates above, the *Pinus-Picea* zone dates mainly from a time when late-glacial conditions prevailed farther north, in New England and the Great Lakes region; the lower part of the zone may date in part from full-glacial time. However, the terms full-glacial, late-glacial, and postglacial have little meaning in an area that was never glaciated. It is likely that no clear distinction can be made between full-glacial and late-glacial time so far from the glacial limit. The entire *Quercus* and *Quercus-Pinus* zones were deposited since the final retreat of ice sheets from the northeastern states.

MIGRATION OF VEGETATION TYPES AND PLANT SPECIES

Much controversy has centered around the extent to which vegetation south of the glacial boundary was affected by the Wisconsin glaciation. Braun (1955) claimed that there were no large-scale changes of vegetation more than a few miles from the glacial limit. Whitehead (1965), in general agreement with Martin (1958), stated that her views were no longer tenable in light of subsequent palynological work. The pollen diagrams from Hack and

Quarles Ponds lend support to the latter view. The site is almost 200 miles from the drift border, yet the lowest pollen assemblage zone suggests vegetation resembling that now found 500 miles to the north. The quantity of pollen of more temperate genera such as *Quercus* in the lowest subzone is small. It is likely that each successive zone and subzone represents vegetation of a more temperate character migrating northward. One must look farther south to find a major full-glacial refuge for deciduous forests.

The changing nature of the zones from south to north, particularly with regard to relative proportions of *Picea*, *Pinus*, and *Tsuga*, suggests that vegetation types do not migrate as coherent units. Rather, it would seem that the component species migrate independently, perhaps at different rates, and that in a new area the components may join together to form a type of vegetation in which the ecological relations have changed to some extent.

It has already been noted that a zone dominated by *Pinus* after one dominated by *Picea* does not occur in these diagrams, just as it does not occur in other diagrams from the southeastern states. Wright (1964) suggested that *Picea* had migrated from a glacial refuge and formed a forest in Minnesota and elsewhere as the glaciers retreated and that *Pinus* had migrated more slowly and replaced *Picea* as the principal component of the vegetation. He suggested that the Appalachian Highlands were the most likely refuge. The fact that *Pinus* and *Picea* pollen occur jointly in large quantity in the lowest zone at Hack and Quarles Ponds tends to support that view. Cushing (1965) points out that a climate unsuitable for *Pinus*, rather than delayed migration, might explain its early absence in the north. Whatever the true explanation, it would seem that the Appalachian Highlands did indeed provide a refuge for *Pinus* and *Picea* at some stage, although they may have had to migrate farther south at the height of glaciation.

REFERENCES CITED

American Commission on Stratigraphic Nomenclature, 1961, Code of stratigraphic nomenclature: Am. Assoc. Petroleum Geologists Bull., v. 45, p. 645–665.

Braun, E. L., 1950, Deciduous forests of eastern North America: Philadelphia, Blakiston Co., 596 p.

—— 1955, The phytogeography of unglaciated eastern United States: Bot. Rev., v. 21, p. 297–375.

Carroll, Gladys, 1943, The use of bryophytic polsters and mats in the study of recent pollen deposition: Am. Jour. Botany, v. 30, p. 361–366.

Cox, D. D., 1968, A late-glacial pollen record from the West Virginia–Maryland border: Castanea, v. 33, p. 137–149.

Cushing, E. J., 1965, Problems in the Quaternary phytogeography of the Great Lakes region, p. 403–416 in Wright, H. E., Jr., and Frey, D. G., *Editors*, The Quaternary of the United States: Princeton, New Jersey, Princeton Univ. Press, 922 p.

Darlington, H. C., 1943, Vegetation and substrate of Cranberry Glades, West Virginia: Bot. Gaz., v. 104, p. 371–393.

Davis, Margaret B., 1958, Three pollen diagrams from central Massachusetts: Am. Jour. Sci., v. 256, p. 540–570.

—— 1960, A late-glacial pollen diagram from Taunton, Massachusetts: Torrey Bot. Club Bull., v. 87, p. 258–270.

Davis, Margaret B., and Goodlett, J. C., 1960, Comparison of the present vegetation with pollen spectra in surface samples from Brownington Pond, Vermont: Ecology, v. 41, p. 346–357.

Deevey, E. S., 1951, Late-glacial and post-glacial pollen diagrams from Maine: Am. Jour. Sci., v. 249, p. 177–207.

Erdtman, G., Berglund, B., and Praglowski, J., 1961, An introduction to a Scandinavian pollen flora: Stockholm, Almqvist and Wiskell, 92 p.

Faegri, Knud, and Iversen, J., 1964, Textbook of pollen analysis: Copenhagen, Munksgaard, 237 p.

Frey, D. G., 1953, Regional aspects of the late glacial and postglacial pollen succession of southeastern North Carolina: Ecol. Mon., v. 23, p. 289–313.

Guilday, J. E., Martin, P. S., and McCrady, A. D., 1964, New Paris No. 4: a Pleistocene cave deposit in Bedford County, Pennsylvania: Nat. Speleol. Soc. Bull., v. 26, p. 121–194.

Hack, J. T., 1965, Geomorphology of the Shenandoah Valley, Virginia and West Virginia, and origin of the residual ore deposits: U.S. Geol. Survey Prof. Paper 484, 83 p.

Harrison, W., Malloy, R. J., Rusnak, G. A., and Terasmae, J., 1965, Possible late-Pleistocene uplift, Chesapeake Bay entrance: Jour. Geology, v. 73, p. 201–229.

Küchler, A. W., 1964, Potential natural vegetation of the conterminous United States: Am. Geog. Soc. Spec. Pub. no. 36.

Martin, P. S., 1958, Taiga-tundra and the full-glacial period in Chester County, Pennsylvania: Am. Jour. Sci., v. 256, p. 470–502.

Ogden, J. G., 1963, The Squibnocket Cliff peat: radiocarbon dates and pollen stratigraphy: Am. Jour. Sci., v. 261, p. 344–353.

—— 1965, Pleistocene pollen records from eastern North America: Bot. Rev., v. 31, p. 481–504.

Whitehead, D. R., 1965, Palynology and Pleistocene phytogeography of unglaciated eastern North America, p. 417–432 in Wright, H. E., Jr., and Frey, D. G., Editors, The Quaternary of the United States: Princeton, New Jersey, Princeton Univ. Press, 922 p.

Whitehead, D. R., and Bentley, D. R., 1963, A post-glacial pollen diagram from southwestern Vermont: Pollen et Spores, v. 5, p. 115–127.

Wright, H. E., Jr., 1964, Aspects of the early post-glacial forest succession in the Great Lakes region: Ecology, v. 45, p. 439–447.

Wright, H. E., Jr., and Patten, H. L., 1963, The pollen sum: Pollen et Spores, v. 5, p. 445–450.

LIMNOLOGICAL RESEARCH CENTER CONTRIBUTION NO. 72
PRESENT ADDRESS: BOTANY DEPARTMENT, TRINITY COLLEGE, DUBLIN, IRELAND

Index

SUBJECT INDEX

AUTHOR INDEX

139